How to teach scientific communication

F Peter Woodford

Chairman of the Council of Biology Editors 1969–70;
Executive Editor of the
Journal of Atherosclerosis Research 1960–63,
Journal of Lipid Research 1964–69,
Proceedings of the National Academy of Sciences, USA 1970–71;
Chief Scientific Officer to the Department of Health, London 1984–93;
Editorial Consultant to the Clinical Research Institute of Montreal and to the Institute of Pharmacological Sciences, Milan

with the collaboration of

Maureen E Goode
Department of Scientific Publications,
M D Anderson Cancer Center, University of Texas

Barbara Gastel
Department of Journalism, Texas A & M University

A COUNCIL OF BIOLOGY EDITORS MANUAL

COUNCIL OF BIOLOGY EDITORS INC

11250 Roger Bacon Drive, Suite 8, Reston, Virginia 20190-5202
Email: cbehdqts@aol.com

PRINTED IN THE UNITED STATES OF AMERICA

LIBRARY OF CONGRESS CATALOGING-IN-PUBLICATION DATA

Woodford, F. Peter, 1930–
How to teach scientific communication / F. Peter Woodford.
p. cm.
Includes bibliographical references and index.
ISBN 0-914340-11-5
1. Communication of technical information. 2. Technical writing.
I. Title.
T10.5.W66 1999
601'.4--dc21 98-54839
CIP

ISBN: 0-914340-11-5

Contents

Preface

This book is a completely rewritten, updated version of *Scientific Writing for Graduate Students*, first published by the Council of Biology Editors in 1968. Like that book, it is aimed at those who wish to teach working scientists how to contribute worthily to the great streams of scientific communication that threaten to cover the earth and, now, to choke the atmosphere. We aim to show practicing scientists that communicating their results and concepts to other scientists and to lay persons is an interesting challenge rather than a dismal chore.

I believe this is best achieved by teaching a course to a class of between 12 and 20 such scientists, both because of the efficiency involved in teaching more than one person at a time and because the interaction with other members of the class as well as with the instructor is stimulating. The book is not aimed at scientists directly, but at the instructor who undertakes to give such a course to however many or few scientists desire it. I also believe that the ideal instructors are themselves working scientists, because they are familiar with the problems and pitfalls not only of publishing but of doing research. However, close to the ideal are those with a scientific background who are not, or are no longer, active in research but are in frequent contact with scientists and with scientific publishing. Such people will, I think, readily accept the main axiom on which this book is based, namely that the subject is most effectively taught by focusing on science, not on language: on the scientific qualities of logic, precision, and clarity of reasoning, not on skill in turning a neat or elegant phrase.

The group of active scientists who come to the class should have produced some research results they wish to publish; the course will then be enriched by class members' interaction with their own data from the very beginning of the publication process. The natural major assignment in the course is to prepare a journal article — or, failing that, a conference poster — based on the scientists' own findings. If any member of the class has insufficient results for this to be a valid option, alternatives are described in Chapter 1.

I trust the postdoctoral fellows and thoughtful scientists up to professorial rank who seek your help will not object to being named here as "students". The term is used for convenience to denote class members, but also with serious intent: those professionals blessed with the greatest maturity regard studying (especially studying for self-improvement) as a lifelong activity, and consider the name a compliment.

To make things easier for the instructor, I have expressed the assignments and the recommended steps as imperatives, so that the instructor can use them unchanged in communicating with the students; the rest of the text addresses the instructor directly.

I hasten to say that the examples of poor scientific writing in Appendix 2, which are provided as model assignments designed to improve the students' scientific style, are not drawn straight from the so-called scientific literature. They are synthetic concoctions, but few readers of this book will fail to recognize solecisms there that are only too familiar in real publications in their field. The subject matter of the examples is drawn from the medical, agricultural, and environmental sciences, but the stylistic principles enunciated here apply to expository writing in chemistry, physics, and engineering just as well.

Throughout the book, however, our major concern is not with weak style or even bad grammar, but with sound logic, good structure, and honesty. A good editor can do wonders with a dull, turgid, verbose paper provided it is clear and well structured, but only the authors can fill in gaps in the logic, clarify the purpose, and supply missing links in a paper that has been ill conceived, badly constructed, or faultily reasoned; and only the author can attest to its scientific integrity. This book's major purpose is to help scientists strengthen their mental powers by preparing their publications in a strictly logical fashion, expressing themselves straightforwardly, and working ethically.

The first nine chapters deal with the writing of journal articles, and the rest with other kinds of scientific communication, still from the standpoint of the instructor: putting together a scientific poster, giving a scientific talk, writing a dissertation, preparing a research grant proposal, and communicating science to the lay public. Most or all of these tasks will fall to the lot of working scientists in the course of their career.

I am greatly indebted to the co-authors of the final chapters — Maureen Goode and Barbara Gastel — for their stimulating input, to David Hayes (Camden History Society, London) for indexing the book, to Peggy Robinson for seeing it through the press with such care and skill, and to Walker & Co, New York, for permission to quote from Dava Sobel's *Longitude*.

F PETER WOODFORD
December 1998

1
Is the author ready to publish?

OBJECTIVES:
To restrain your students from prematurely attempting publication.

To emphasize that the course is about scientific communication, not stylish language.

To discuss the roles of co-authors.

To introduce the concept of publication ethics.

YOU SHOULD READ:
Maeve O'Connor. *Writing successfully in science.* London and New York: Chapman & Hall; 1991.

Council of Biology Editors Editorial Policy Committee. *Ethics and policy in scientific publication.* Bethesda, (MD): Council of Biology Editors, Inc; 1990.

Same committee. Redundant publication. *CBE Views* 1996;19:76-7.

Thomas A Lang and Michelle Secic. *How to report statistics in medicine: annotated guidelines for authors, editors, and reviewers.* Philadelphia, (PA): American College of Physicians; 1997.

The references at the end of this chapter.

YOU WILL NEED:
Copies of Table 1.1 (p 9).

ASSIGNMENT:
Draw up tables and figures complete with titles, captions, and footnotes, with the aim of defining the content and magnitude of the proposed major assignment.

TIMING:
1–2 hours for this opening session.

Your course will succeed best if it includes graduate students with at least a year's experience in research, together with mature scientists who, scarred by earlier battles with journal editors, wish to ease the pain of writing for publication. If you have not a class but only one scientist to teach, the principles still apply, but then the relatively inexperienced will not benefit from the rueful confirmation, by more experienced scientists in the class, of some of the points you will be making.

The scientists who have come to you for help may be hoping for a "quick fix" to their writing skills that will hurtle them into print within weeks. Begin by restraining the hotheads in the group who imagine that with the aid of a magic formula they will effortlessly sail into publishing an article in a major journal to the sound of mellow plaudits all round. Tell them that the road to being a good writer of scientific articles is a long, but not necessarily a hard, one. The secret is to take the process calmly and steadily, in a sequence that is not often advocated in books on scientific writing, but which I have found has the most appeal for scientists.

The course described in this book (Chapters 1–9) on preparing and publishing a journal article is based on a series of 22 steps (Table 1.1). This may seem at first a daunting number. Your first task is to convince your class that

- the sequence is logical,
- it concentrates on data rather than words,
- it makes the process more efficient because it averts the need for endless rewriting, and therefore
- it actually shortens the process.

This first session takes the scientists in your class through a discussion of the first four steps in Table 1.1. The remaining sessions, perhaps 10 or 12 in all, will cover the rest in rational sequence, with minor assignments along the way, culminating in Step 21: submission to a journal of a completed article, preparation of which is the major assignment tackled throughout the course. Step 22 suggests the best way of responding to the editor's decision letter.

Distribute copies of Table 1.1 and begin discussion of the steps.

Note that Table 1.1, and the wording of the steps and assignments throughout the rest of this book, are addressed to the students, whereas the text around them is addressed to you, the instructor. In the table, the steps, and the assignments "you", "yourself", and so on speak to the student; the same words in my text are addressed to you, the instructor.

Table 1.1 Writing a good journal article

Step 1	Ask yourself whether the time is right
Step 2	Clarify your conclusions by preparing tables and figures complete with titles and footnotes
Step 3	Decide who will be co-authors
Step 4	Consider the ethics of scientific publication
Step 5	Relate your conclusions to the existing body of knowledge
Step 6	Write a working title and abstract
Step 7	Choose the target journal and make notes on its Instructions to Authors
Step 8	Name the main sections
Step 9	Fill section files with relevant brief notes in any order, to form "ragbags"
Step 10	Arrange the contents of the ragbags logically
Step 11	Finalize the design and content of tables and figures
Step 12	Make a topic outline and consider writing a sentence outline
Step 13	Write the first draft continuously, collecting references as you go
Step 14	See if the first draft needs major alterations
Step 15	Have the illustrations prepared in final form for the target journal
Step 16	Polish the prose
Step 17	Rewrite the title, and structure the abstract
Step 18	Request private review by three independent critics and your co-authors
Step 19	Re-read the Instructions to Authors and make any necessary adjustments
Step 20	Revise as many times as necessary
Step 21	Submit the article to the journal
Step 22	Analyze the editor's decision letter and respond appropriately

Step 1 Ask yourself whether the time is right

Discussion of this step enables you, the instructor, to concentrate on the research findings that your students want to publish. (Throughout this book, "students" include all those scientists, including mature and experienced ones, who have enrolled in your course or who have approached you for individual tuition in the production of publishable articles. Explain that the term is used for convenience, but also because the greatest scientists have thought of themselves as students acquiring knowledge all their lives, so they should not take offence.)

The only, repeat only, justification for publishing a scientific article is to present significant new data or concepts. Focus on the data your students have produced and on whether the time is right to make them public. Such focusing will

- establish that your primary concern is with science rather than the vagaries of language
- secure their confidence in you as a colleague who is, as it were, playing in the same team, not an outsider carping from the touchline
- relieve students of their fear, mild in some but powerfully strong in others, of being woefully inadequate in stringing words together — because you will be talking exclusively about numbers and images, not words, in this session.

The supremely important feature of a good scientific paper is logical reasoning. Emphasize, then, that in writing a scientific paper your students will be playing to their strengths. Their training has been directed toward logical thinking, scientific deduction, and rigorous analysis of possibilities and certainties. All they have to do is to apply this training to the task in hand.

Ask them why they embarked on this piece of research: What were its antecedents, either in their own previous work or in the literature? What was the logic that led them to conduct these experiments or make these observations, and what logical thread leads them to their conclusions? In this connection, require that they ask themselves some searching questions *and put the answers down in writing.*

- What is the over-all falsifiable (null) hypothesis or research question?
- Did the experiments really test it, statistically and otherwise?
- Were there subsidiary hypotheses, subject to the same criteria?
- What was the experimental design, why was it chosen, should its particular strengths be described, are there any potential criticisms of it that should be admitted or refuted?
- Were the controls that were used appropriate and sufficient?

- Did the author consult one or more statisticians about the experimental design as well as the validity of the conclusions?
- And how significant are those conclusions in terms of human welfare or of pure knowledge?

This last question reveals whether the student is in fact ready to publish. If the conclusions have no clear significance, it is time to return to the bench or the field study and either get further evidence or change tack. If any of your students evince the attitude that they simply must publish something, somewhere, in order to advance their career, tell them of the dire negative effect that prematurely published work (always assuming that it *gets* published) will have in the future when they are struggling to get a research grant application approved.

But if there *is* a meaningful contribution to be made, of specialist or general importance, it is time to publish. Already, ideas about where to place the eventual article will be burgeoning; however, you will need to rein your most eager students back, and have them embark first on Step 2.

Step 2 Clarify your conclusions by preparing tables and figures complete with titles and footnotes

Now home in on the conclusions even more closely. In the eventual article, the data will be presented in tables and figures, and if these are well designed they can tell an experienced reader the whole story, including the rationale underlying the investigation and the bulk of the conclusions, without any reference to the text. So this is where preparation of a paper should begin. Show your students how vital it is to put much thought into

- the titles of tables and headings of columns
- the provision of accurate wording in footnotes to tables
- the choice of tables versus graphs, pie charts, histograms, or bar charts to display numerical results
- the choice of axes, labeling, and titling of graphs and other figures
- selection of the field and degree of magnification of any photographs,

all in the service of conveying as much accurate information as possible to the eventual reader *through these means alone.*

Drawing up tables and figures in complete form can reveal devastating gaps in the evidence, and then more research is needed before any writing is contemplated. Even if all is well, the exercise will fulfill the *private* purpose of clarifying the author's thoughts. (You will be showing at Step 11 how to mold the tables and figures so as to serve the *public* purpose of informing readers.)

ASSIGNMENT
Before the next session, hand-draw tables and figures complete with surrounding wording, and think about the type of journal article to which they should lead.

Before going any further, authors must now check that they have used the right kind of statistical analysis, and that they know how they are going to report the results and interpret them. This means consulting the best and most recent book on statistics and statistical reporting, which in 1998 was *How to Report Statistics in Medicine* (see head of this chapter), most of whose definitive and clear guidance is applicable to all branches of science.

If these processes indicate that the would-be author is not yet ready to publish, and if the problem cannot be resolved by reconsidering or recalculating the data, there is no alternative for him or her but to resume the investigation and postpone preparing the article until the gaps have been filled. However, students who find themselves in this pickle need not give up your course. They can learn a lot from it even if they do not have a final set of data to work from — not only about writing, but about scientific reasoning in general.

It is to be hoped that this situation will not arise often, as the course works best if the major assignment is the preparation of a real article for publication; but students with inadequate data can often find a *portion* of their results that can form the basis of their major assignment, presented as a Short Report or Methodological Note. Every publication, however small, needs to go through the whole process and the 22 steps.

Notes on the major assignment

If some of your students really do not have new results of their own ready for writing up as a major assignment, try suggesting a few alternatives:

- They may be partway through an investigation and know what conclusions they will draw if the results go one way or another: writing the article on the basis of predicted results will be quite as instructive in logical thinking and its expression as if the results had already been obtained. Needless to say, the parts of the eventual paper that are fictitious should be clearly labeled as such at the top and bottom of the relevant page(s), and the whole should never be submitted to a journal!
- Alternatively, a set of data that is incomplete and would not warrant a full journal article might well be built into a conference poster and its corresponding abstract, since these often describe work in progress that is not yet publishable. This too could serve as a major assignment. The same

steps must be followed as for a journal article, with the slight modifications described in Chapter 10.

- Students who have published previously could rewrite an earlier paper, using the process presented here instead of the one originally used.
- As a final option, a student could write a purely theoretical paper, perhaps in the form of a research proposal to an imaginary, hardheaded board of directors with limited scientific knowledge.

Step 3 Decide who will be co-authors

Until now, we have been assuming that each of your students will be sole author of the article. This is, nowadays, seldom if ever the case. The tendency toward longer and longer bylines, even in the life sciences, is one that journal editors view with dismay, finding it hard to believe that a dozen or more scientists have been closely involved with every aspect of a minor methodological improvement or with a paper that adds a tiny grain of sand to the mountain of previous scientific knowledge.

Recent discussions in the USA and elsewhere have concluded that it is best to decide who will co-author the eventual paper(s) even before a study is begun, and when the study is of limited scope, with few workers, all of them active, this step may pose no difficulties. But when the article to be published forms part of a large complex of mutually dependent studies, life may not be so easy. However, you may give the following simple guidelines.

Dissuade your students from automatically adding the name of the distinguished but remote head of department whom they may meet socially once a year. Nor should they, in a kind but misguided gesture, add that of the indispensable but non-scientific laboratory cleaner and glassware purifier. The most widely accepted guidelines[1, 2] hold that authors named at the head of a paper must have been crucially responsible for

- the design of the investigation,
- managing the conduct of the experiments or observations,
- interpreting the results and shaping the conclusions,
- writing, reviewing, and revising the paper, or
- all of these.

With rare exceptions, names of the following should appear not in the by-line but in the Acknowledgments section: technicians following instructions, research students gathering some of the data, radiographers and other paramedical staff obtaining data on patients as part of their ordinary work, statistical advisors, secretaries, editorial assistants, coding clerks, and lenders or providers of special materials or equipment.

Very recently[3–5] some important medical journals have replaced authorship with "contributorship". Contributors fulfill the same criteria as authors, but the nature of their contribution is defined either in a footnote or at the end of the paper, and some journals may aim to have the proportions of contribution approximately quantified. In addition there is a "guarantor", who by assuming this title personally guarantees that the research has been properly and ethically conducted and that the findings have been reported honestly and accurately.

The new system is intended to make the distinction between two main traditional roles of authors: taking credit for adding to the sum of human knowledge, and assuming responsibility for its accuracy and truth. Whether the system will succeed, only time will tell; it is being introduced gradually, in some journals only if the authors/contributors wish it. As far as the rest of this step is concerned, the same principles of cooperative discussion apply either to co-authors or to contributors, and I shall for simplicity continue to call them authors.

Identifying the co-authors of the prospective paper at an early stage is important not only in preventing later disagreements, but in helping to define the scope of the paper. Often, members of the same large team will have been engaged in overlapping investigations, some of which will come to fruition earlier than others; once the authors of the prospective paper have been identified, they can collectively decide the exact subject of *this* paper and therefore which results to include and which to exclude. In the course of this discussion the group will depute one person to prepare the first draft, right up to Step 18. This is the person we call "the student" in all that follows. Although he or she will take the lead in the preparation and writing steps, this of course does not preclude consultation with one or more of the co-authors along the way.

You (or your students) may wish to deal now with the question of the order of authors' names on the title page. This issue is addressed differently in different institutions. The most logical and scientific way would be to list the authors in descending order of the magnitude of their contribution (conception and design, execution, or interpretation) to the work reported. But other conventions exist too, such as that the head of the group and instigator of the work is named last. There is no unequivocally right, and fair, way of handling the matter. Suggest that the co-authors agree a provisional order of names at the very start of planning the writing, with the opportunity of reviewing the order at Step 18 when the article is in nearly final form.

Step 4 Consider the ethics of scientific publication

The co-authors must come to their first meeting keenly aware of the ethics of scientific publication. They will, at a much earlier stage, have taken cognizance of the rules regarding ethics of experimentation with respect to the welfare of animals, people, and the environment. The student must remember to insert into the article a statement on how those rules were followed. However, it is just as important, if we are to avoid pollution of the stream of scientific information, to avoid certain actions that may be tempting to the novice — and even to the experienced scientist who is over-ambitious or naive. The actions all involve over-publishing. Some scientists don't know that it is unethical

- to publish the same data in different forms in different journals,
- to submit the same paper simultaneously to more than one journal with the intention of withdrawing it from the journal that responds last,
- to carve up a single investigation into several small articles, or
- to publish partial results of a single investigation at intervals, in a series of progress reports.

Explain why any of these actions can be confusing as well as burdensome to readers (including reviewers and editors), can give rise to apparent independent confirmations of initial reports, and can unbalance meta-analyses. Explain also how they increase the cost to editors, reviewers, and the publication system as a whole. And point out that penalties can be severe if such "redundant publication" is detected by those editors who punish it by publishing the fact, identifying the authors concerned, and putting them on at least a few important blacklists. In teaching this part of the course, you may want to refresh your memory about journal editors' attitudes by reading the article on ethics in publication cited at the head of this chapter.

References

1 Appendix on "Authorship" in Huth E J. *How to write and publish papers in the medical sciences.* 2nd ed. Baltimore (MD): Williams & Wilkins; 1990. See Annotated Bibliography.

2 Information for authors. *Ann Intern Med* 1998;129(1):I-4–I-12 (and subsequent versions).

3 Rennie D, Yank V, Emanuel L. When authorship fails: a proposal to make contributors accountable. *J Am Med Assoc* 1997;278:579-85.

4 Horton R. The signature of responsibility. *Lancet* 1997;350:5-6.

5 Smith R. Authorship is dying: long live contributorship. *Br Med J* 1997;315:696.

Summary of Chapter 1

Vital preliminaries, with a long hard look at the results

- Restrain students from premature publication (Step 1).
- Start from students' data, not consideration of language.
- Present the table "Writing a good journal article" listing 22 steps for the process.
- Stress that the prime concern is logical reasoning.
- Discuss the presentation of results *and the logic linking them to previous knowledge* through the medium of well-chosen and well-constructed tables and figures (Step 2).
- Reveal how Step 2 determines whether the work is ready for publication.
- Discuss the course's major assignment.
- Guide the choice of co-authors (Step 3).
- Identify your students as "preparers of the first draft".
- Discuss the ethics of publication (Step 4) and some of the penalties for being unaware of or disregarding them.

2
Choosing the target journal

OBJECTIVES:
To prevent authors from wasting their time in directing their efforts to an inappropriate journal.

To introduce students to the discipline of defining purpose and scope at the outset and matching these to a target journal's stated purpose and scope.

YOU WILL NEED:
Copies of leading journals, some of wide scope (e.g. *Science, Nature, Proceedings of the National Academy of Sciences, USA*), some of greater specialization (any high-quality, peer-reviewed specialty journals appropriate to your students' disciplines). Choose the issues containing the Instructions to Authors and provide photocopies of these for each journal.

One or two copies of *Current Contents* (Institute for Scientific Information, Philadelphia, PA).

A volume or copies of a page of *Science Citation Index.*

TO READ FOR YOUR AMUSEMENT:
F Peter Woodford. Sounder thinking through clearer writing. *Science* 1967;156:743-5.

ASSIGNMENTS:
Briefly abstract a published paper.

Write a working title and abstract for the major assignment.

Select a target journal; make notes on its Instructions to Authors.

TIMING:
1–2 hours, even though you cover only three steps.

In the second session we move on to some actual writing — still not writing for the public, yet, but writing privately to clarify thought.

It is often said that clear writing can result only from clear thinking; the converse is equally true, though rarely stated. Just as (see Chapter 1) preparing meaningful titles, footnotes to tables, and legends to figures helps to clarify the purpose and scope of the investigations as well as their logical links to each other, writing important thoughts down and then examining the result invariably clarifies the thinking.

The value of pinning things down in words becomes abundantly clear in the next step.

Step 5 Relate your conclusions to the existing body of knowledge

There must have been a reason for the student to embark on the work to be described, and that reason lies in previous publications. Get your students to ask not only "What are my conclusions?" but "What was my question, and why did I ask it?" — *and to record the answers*, whether they use a pen, pencil, or word processor.

Recording the answers to these questions may seem simple to many of your students, and your insistence on it almost trivial. But I guarantee that a large majority will find the outcome surprising: they will discover that they are not as sure of the details and arguments in previous work as they thought, and that their conclusions have to be expressed more cautiously than they had hoped. The process may even, in the worst scenario, signal that the work is not after all ready for publication — in which case no time will have been wasted in writing a paper with small chance of acceptance. But, as suggested in Chapter 1, the student can still be encouraged to complete the course, citing theoretically possible results from well-designed further experiments or observation.

A useful exercise at this point, which may help to drive the point home and will certainly help students to read others' papers more critically, follows.

> ASSIGNMENT
> Ask "What question has been asked, what is the evidence, and what are the conclusions?" of any paper in your current reading. Ignore the published title and abstract at first, and write your answers to these three questions from a reading of, first, the tables and figures, and then the whole text. Then compare what you have written with what the author has claimed in the title and abstract.

One more step, and your students will be ready to choose their target journal.

Step 6 Write a working title and abstract

After completing Step 5, your students should have all the elements they need for precise definition of the coverage, the boundaries, and the limitations of the paper they are preparing. Now they should crystallize these thoughts by formulating them in words, in a form close to what will eventually be published as a title and abstract.

Many inexperienced writers plunge into the detail of their paper, possibly starting with Methods as the easiest section, without knowing whither the article is going to wander, thereby producing the kind of paper that is most irritating to reviewers and most likely to be rejected. Even if each sentence is perfectly constructed and on its own makes sense, readers will be unable to see the point of the whole and will ask "What on earth are the authors driving at?"

Some of your students may have been told by other mentors to leave the construction of title and abstract to the end, since only when the paper is finished will they have a clear idea of what it is about.

Dismiss any suggestion of this sort. Although the emphasis, or some details, in the working abstract may need to be changed in the light of later thinking, and an even more precise and informative title may eventually come to mind, it is vital that the author know *now* what the paper will set out to say. Stress the word "working" in this step, and point out that formulating a working title and abstract is a clever use of writing to clarify thinking.

> ASSIGNMENT
> Write a working title and abstract of the projected article on the basis of the tables and figures prepared at Step 2 and the questions and answers in Step 5. Examine them critically and adjust them if necessary, as they will be your constant guide throughout the rest of the major assignment.

Step 7 Choose the target journal and make notes on its Instructions to Authors

For some students, choosing a journal to which the paper will eventually be submitted will appear easy: most of their reading may be in a few journals, and those are the ones they think of first. But there are many factors to be considered in choosing a journal from among the myriad available. The prime factor must be matching the purpose and scope of the article, as described in the working title and abstract, to the statement of Purpose and

Scope in the Instructions to Authors of the potential targets.

Not all readers of journals know of the existence of Instructions to Authors (or Guidance for Contributors) for every journal; even if they do, they may not think of reading them at this stage. It is natural to think that the instructions merely specify things like the width of the margins, the number of copies to be submitted, and the kind of diskette and software to be sent together with the hard copy. But now, not later, is actually the time to look at the passage of the Instructions to Authors that specifies the purpose and scope of the journal, since these govern the editor and reviewers of the journal in their consideration of whether submitted papers are suitable for it or not.

Of course, no one is going to submit a report of engineering research to the *Journal of Biological Chemistry*, but the statement of Purpose and Scope in the potential target journals may indicate rather more subtle distinctions that may sway the would-be author's choice. For example, the journal may exclude purely methodological papers, or those involving animal experimentation because it records only clinical research. The statement of Purpose and Scope will reveal whether the journal welcomes or rejects brief communications or letters and whether there are limitations as to the length of the paper, the number of references, the number of photomicrographs, or the use of color. Above all, it will clearly set out the extent of the subject matter the journal aims to cover.

If these matters match up with the working title and abstract in two or more apparently suitable journals, authors should move on to other criteria:

- How widely read is the journal: what is its circulation, is it international, does the journal's Contents page appear in *Current Contents*, is it abstracted by major online searching services, what is its impact factor derived from the *Science Citation Index*?
- How respected is the journal: do the names of the members of the editorial board give a clue here, can the students ask their supervisors or more experienced colleagues?
- Is one more likely to get helpful, knowledgeable reviewers' comments from this journal or that?
- How quickly does the journal publish (deduced from the published dates of submission and publication)? Is this vitally important in this particular case?
- What is the quality of the halftones? Is this important for the article?
- Is the journal published in electronic form available on the Internet, and if so is there a hard-copy version as well?

Encourage students to look beyond their usual reading range by scanning the pages of the relevant edition of *Current Contents.* The titles of articles listed there will often suggest the potential suitability of several different journals, which must then be examined further, beginning with the statement of Purpose and Scope.

Strongly discourage students from postponing their choice of journal until after they have prepared the paper. Good communicators are constantly aware of their audience. Over-anxiously spelling out detail will annoy a specialized audience; omitting necessary explanation for a broader readership will baffle and confuse. Keeping in mind the probable types of reader of the chosen journal is the best way to avoid both these faults.

Recently, some journals have opted to publish not on paper but electronically, on the World Wide Web. While this seems highly attractive at first sight, especially in the hope of sidestepping the usual several months' delay between acceptance and printing, there are dangers and complications here that you should explain to your students. Among them are the proposals to open up published work to interaction with Web browsers. I don't describe the complications because the situation and the relevant policies and copyright questions are changing so rapidly that anything I say will be out of date before this book reaches the printer. Over to you to become well informed!

Now insist that, once students have chosen their target journal, after thorough examination of the Purpose and Scope and the journal itself, they read through the rest of the Instructions to Authors and note down any points they must bear in mind during subsequent preparation of their papers. Many of these have been standardized, for medical journals at least, by an international committee of journal editors.[1] But each journal has its idiosyncrasies, and although the student can safely relegate to a later stage such details as the number of copies to be submitted and where, exactly, the page number should appear on the page, it *is* important to know whether articles must be structured in a certain way; whether "personal communications" must be accompanied by the originator's written agreement; whether titles of articles and inclusive pagination must be given in references; and especially whether there are other specific conditions peculiar to this journal. None of these cause much trouble if attended to during the writing, but any one of them will cause a great deal of trouble if discovered just before submission.

ASSIGNMENT
Select possible suitable target journals for the major assignment, choose one with the aid of its statement of Purpose and Scope and using other

criteria, after consultation with co-authors or mentors if appropriate. Make written notes on its Instructions to Authors for guidance during further preparation of the paper.

Reference

1 International Committee of Medical Journal Editors. Uniform requirements for manuscripts submitted to biomedical journals [5th version]. *Ann Intern Med* 1997;126:36-47 (and other journals).

Summary of Chapter 2

Writing the working title and abstract so as to make the right choice of target journal

- Teach students to clarify their thoughts, in particular the purpose and scope of the proposed article, by writing them down.
- Recommend (Step 5) that students ask themselves "Why did I/we do this work?", "What conclusions do we draw?", "What is their magnitude and significance?" and thereby define the type and scope of the proposed article.
- Advocate writing a *working* title and abstract (Step 6) now, not later.
- Advise students to choose the target journal (Step 7) now, not later, by matching the outcome of Step 6 to the journal's stated purpose and scope and using other criteria.
- Have students record notes on policies, conditions, and restrictions stated in the target journal's Instructions to Authors.

3
Marshaling facts and ideas into section files

OBJECTIVE:
To overcome students' "writers' block" by teaching them to assemble their material pell-mell into "ragbags" corresponding to the main sections of a journal article, and sequencing the items only later.

YOU WILL NEED:
Copies of the Instructions to Authors for a few journals of wide and narrow scope.

YOU MAY NEED:
A selected list of books on scientific writing in different specialized fields for those students who wish to study the subject for themselves alongside your course (see Annotated Bibliography, p 185). You should stress, however, that they will not find the approach you have used up to now in any of them; most start with drafting an outline, none with a hard look at the data and the application of a logical scientific approach.

Particularly useful for summarizing the principles in this chapter in different words is *Communication for Engineering Students,* by John W Davies (Longman), Chapter 5, "The writing process".

For students who are involved in reporting clinical trials, copies of the checklist for reporting such trials, on p 638 of *J Am Med Assoc* 1996;276:637-9 and p 742 of *Ann Intern Med* 1996;124:741-3.

ASSIGNMENT:
Assemble into section files items to be included in the major assignment, according to fixed criteria but in any order, so as to form ragbags. Then, put the items in each ragbag into logical order, re-allocating items into a different section if necessary.

TIMING:
1 hour for teaching. The assignment is a lengthy one.

Step 8 Name the main sections

The Instructions to Authors in the target journal may stipulate that all articles must use the IMRAD structure — Introduction, Methods (or Methods and Materials), Results And Discussion — in that order. This conventional form for scientific articles has emerged in the 20th century as the most convenient to readers, and they are used to it, so advise students to adopt it if possible. But sometimes, for instance in a purely theoretical or conceptual piece, and certainly in a review article, it would be quite inappropriate. Moreover, some journals (e.g. *Nature* and some specialized journals) require their own distinctive structure. So encourage your students to give some thought to whether they are embarking on a straightforward research report or whether they require a different kind of plan. However, if the preliminary decision is against IMRAD, students must be very clear why they have abandoned it; what, precisely, are the merits of the new one; and which journal is amenable to the chosen form.

You may, if you have time, want to discuss with students why one earlier form, in which the Methods section came at the end under the heading "Experimental", wilted and died for most journals (presumably when it was realized that the experimental conditions critically affect the results and should therefore precede them), but survives in a few. You could also point out that certain journals, including *Nature*, ask that experimental details be confined to the captions to figures and tables.

Whatever structure is adopted, there will be major sections characterized by main headings such as Introduction, Methods, or some other type of main heading. The headings of the sections will govern what parts of the author's material is put into each of them. For a highly complex journal article, students may wish to think, even at this early stage, about formulating subheadings within the main headings. These will act as signposts — for the author as well as the reader — showing the direction in which the article is heading at any point. More usually, however, appropriate subheadings emerge after the next two steps.

Step 9 Fill section files with relevant brief notes, in any order, to form "ragbags"

This step particularly helps scientists with "writer's block" — those who seize up at the sight of a blank sheet of paper or screen — because it asks them to make notes and sort them instead of having to write complete sentences before they know where they are going.

Recommend that each student labels separate sheets of paper, or word-processing files, **Introduction**, **Methods**, **Results**, **Discussion**, or other main section headings. He or she should then put into each section file, in brief personal shorthand — *not* polished sentences — all data, concepts, and vital bibliographic references that belong there. No thought of logical order disturbs the rush of notes onto the page or screen; all the writer considers is in which section the fact or idea properly belongs. It is as though one were unloading furniture from a removal van into the empty dwelling that is the future journal article, thinking only whether the item belongs in the kitchen or the bedroom and not where to place it within that room.

Notes for the article will inevitably come to mind in chaotic fashion as the author combs through the relevant laboratory or field-trip notebooks, reprints, file cards, and printouts from literature searches, and the notes should be added to the appropriate "room" in similarly chaotic fashion. If the student is engaged in reporting a clinical trial or part of one, the checklists referred to at the head of this chapter can be used to help. More generally, the headings suggested by the target journal for structured abstracts may be similarly used to check for inclusiveness.

In addition to sheets or computer files bearing the titles of the sections, there must be one bearing the title **Acknowledgments** (into which a note of any special services rendered during the investigation will go) and one headed **References**. There may also be one labeled **Ethical considerations** (usually within **Methods**) if this was a major issue in the work.

For each item under consideration for entering into a section file, the article writer must decide, *with frequent reference to the working title and abstract*:

(a) whether it is relevant and absolutely must be included, or whether it should be thrown out in the interests of clarity, and

(b) if it *is* essential, under which heading it should go.

The decisions under (a) help to keep the Introduction and Discussion short, crisp, and relevant; those under (b) help to keep Results and Discussion sections distinct. One last question then needs to be asked: Has every item that should be present been included? Outlying results, for example, that detract from a particular interpretation may not be omitted for that reason alone.

The distinctive feature of this procedure is that items are added *in any order*. There is no shame in having one's ideas tangled like a twist of DNA in one's head — that's what aids the lateral thinking that so often leads to real scientific advances. But it can be disastrous to start writing for others' consumption before one has teased out one's thoughts into nice continuous lines. Another metaphor the instructor can use is that each section constitutes a ragbag at this stage, a bag into which various items are thrown pell-mell and from which they will later be selected to form an attractive and meaningful pattern.

This process has been found useful not only by those with writer's block but also by those who write with ease, because it defines and delimits the material that is to go into the article as a whole. It also ensures that each piece of information goes into the logically correct part of the article and is not used twice.

Step 10 Arrange the contents of the ragbags logically

When the author is satisfied that everything is present that should be included, that items of peripheral relevance have been excluded, and that each item has been allocated to the most appropriate section, he or she must then examine the resultant ragbag of bits, relate the items to one another, and put them in sensible sequence. The main line of reasoning, with its attendant tributaries, usually begins to emerge as if by magic. So does the wording of possible subheadings, which may or may not be used in the eventual article but which can help to guide the author in writing the first draft.

Whether the section files should consist of sheets of paper or computer files is a matter of taste. The first can be more conveniently carried around from laboratory or field station to library; the second may be easier to put into logical order when the time is ripe, without the need for the swooping lines and arrows that were indispensable in the ink and paper era.

ASSIGNMENT

Allocate to a small number of section files, which must include two labeled **Acknowledgments** and **References**, the items to be included in your major assignment — in abbreviated note form and in no particular order. Include all necessary items, and ensure they are relevant and appropriately allocated. Then put the resultant ragbags into logical order, feeling free to re-allocate items if the original placement seems wrong.

Summary of Chapter 3

Inclusive stocking of section files to form ragbags, followed by logical marshaling of the contents

- Point out that the usual order of sections is IMRAD (Step 8), and there must be good reasons to use a different one.

- Teach the use of chosen main headings (e.g., Introduction, Methods, Ethics, Results, Discussion or Conclusions, Acknowledgments, References) to label section files, whether on sheets of paper or in a word-processing program.

- Advise students to stock the sections in random order with appropriate items in brief note form (Step 9), asking in the light of the working title and abstract constructed at Step 6:
  Is this item relevant? (If not, delete it.)
  If so, in which section should it go?
  Have I included everything necessary?

- Only after this, arrange items in logical sequence (Step 10).

4
Designing tables and figures

OBJECTIVE:
To demonstrate that illustrations are powerful modes of communication in themselves if they are logically designed, titled, footnoted, and captioned.

YOU SHOULD CONSULT:
Council of Biology Editors Scientific Illustration Committee. *Illustrating science: standards for publication.* Bethesda (MD): Council of Biology Editors, Inc; 1988.
This book is chiefly aimed at publishers and printers, but Chapter IV usefully illustrates the different kinds of graphical demonstration of data points available to authors.

Mary Helen Briscoe. *A researcher's guide to scientific and medical illustrations.* New York: Springer-Verlag; 1990.
This book gives instances where author and illustrator need to communicate to produce the most telling pictures.

YOU WILL NEED:
A list of titles of successive tables demonstrating redundancy and inconsistency, for example that on p 30.

Some badly constructed tables, either early examples from Appendix 2 or derived from your own observation.

ASSIGNMENTS:
Eliminate redundancy and inconsistency from a series of fictitious titles of successive tables.

Improve a badly designed table and compose the corresponding part of the text.

TIMING:
1 hour, plus discussion in a subsequent class.

Step 11 Finalize the design and content of tables and figures

The process of stocking the section files pell-mell and then logically arranging the resultant ragbags (Chapter 3) will have bolstered your students' confidence and crystallized their vision of what their article will cover — and what it will not. Now the students should look again at the data included in the tables and figures sketched at Step 2. Do the design and wording around these illustrations fit the concepts and the structure that have been developed as the section files were sequenced? Within which part of the Results should they come, and in what order? Should some of the tables be split or, on the contrary, combined for best communication? As part of this process, the author should quickly check the target journal's preferences with respect to the construction of tables and their captions. Thus, most figure legends start with a phrase identifying the figure's subject, followed by complete sentences giving experimental details or keys to symbols used in the figure; but in some journals, especially in the basic sciences, the legend begins with a sentence stating what the authors believe the figure demonstrates.

Ideally, the tables and figures on their own should tell virtually the whole story to a knowledgeable reader who has read the article's title and abstract. A good test of whether they do so is to take the titles given to the illustrations and list them separately, in the order planned. This will quickly reveal any unnecessary repetition, internal contradiction, and lack of clarity of intention. It will also strengthen the sense of structure that is so important to bear in mind when the article is being written. And it will show up any gaps in the information provided in and around the illustration: it should never be necessary to add the words "see text" to an illustration.

Enriching the information conveyed by the illustrations has the added important advantage of helping to shorten the text, which will then need to refer only in the briefest terms to the design of the experiments or patterns of observation and can concentrate on the conclusions and interpretations to be drawn from them. The text should never, ever, repeat the values shown clearly in a table or accurately indicated by a graph or histogram — and there will be no temptation to do so when the author is confident that the illustration lays it all out in the clearest and most striking manner.

Drive home these points by using the assignment on table construction and titling that follows.

ASSIGNMENT
Eliminate redundancy and inconsistency from the following titles of successive tables:
Table 1. Time course of hydrolysis of glyceryl 1-monoacetate by rat liver mitochondria.
Table 2. Effect of suspension medium on hydrolysis of 3-acetoxypropane-1,2-diol as measured by UV absorption.
Table 3. Effect on monoacetin hydrolysis of activators and inhibitors.
Table 4. Effect of substrate concentration on hydrolysis by rat liver mitochondria.
The instructor may need to explain that the three different chemical names in the titles of Tables 1–3 are exact synonyms. An improved list of the above titles would be as follows:
Table 1. Time course of hydrolysis of 1-monoacetin by rat liver mitochondria.
Table 2. Effect of suspension medium.
Table 3. Effect of activators and inhibitors.
Table 4. Effect of substrate concentration.

This careful examination of the tables and figures in conjunction with the content of the sequenced section files may have an unexpected adverse effect. When the suggested conclusions are matched to the facts shown in tables and figures, it may be seen with dismay that they are not, or not fully, supported. The choice is then either to rein in the conclusions or to abandon the article until more results are obtained.

The sound construction of tables and figures is so important, and so central to students' interests, that it is worth setting a further assignment on it.

ASSIGNMENT
Improve a badly designed table and write the part of the text that refers to it.

The table (or tables) you give to students to improve should bear (a) an uninformative title, (b) unnecessary repetition of words or phrases, (c) too many columns, and (d) inadequate footnotes. You can also include "accompanying text" that repeats in nauseating detail all the information and data in the table without any additional information or interpretation, or repeats the information with the kind of "elegant variation" that makes it doubly difficult to read the table. Examples of relatively simple, and more complex, tables are given in Appendix 2, but you may wish to construct your own. Be sure to discuss the improvements in a subsequent class, and hand out an improved version.

The points to make to the members of your class with respect to their major assignment are as follows. Whether you give examples before they hand in their assignment or afterward is up to you.

- Design each table and figure as a complete unit of information.
- Write informative titles for tables and figures, always with the intention of revealing the *purpose* of the illustration, thus using words like "increase in" or "specificity of" rather than "comparison of" or "values for".
- Construct tables logically, generally with a smaller number of columns than rows, with control values first and values to be compared in adjacent columns or rows; and use internal headings to simplify reading, comparison of values, and immediate grasp of significance.
- Do not alternate arbitrarily between tables, line graphs, pie charts, and histograms, but decide which form conveys the information most clearly (note in passing that histograms are preferable to line graphs if a continuous curve cannot legitimately be drawn between data).
- On line graphs, show all data points, using conventional symbols and with vertical lines representing SD or SEM (standard deviation or standard error of the mean), *stating which of them is represented.*
- Choose appropriate units for labeling the axes of graphs.
- Never extrapolate beyond the data.
- Work hard at writing brief headings to columns in tables, supplementing by footnotes if necessary.
- Use footnotes to tables, and legends to figures, to explain all abbreviations and statistical calculations.
- Footnotes and legends must state whether any variation shown is SD or SEM; the corresponding number of observations, *n*, *must* be given somewhere, whether in the body of the table, a footnote, or the legend.
- Decide whether instrumental tracings on paper or gel autoradiographs actually need to be shown or are better described in the text.
- Draw up a list of titles of tables and figures for the whole article, eliminate unnecessary repetition, and ensure that titles relate to one another, i.e. are consistent in terminology and parallel in form (see p 30, also pp 138–139, for examples of bad and better sets of titles).
- Simplify the first attempts at tables and graphs to the utmost extent consistent with conveying the message, the process of simplification to include combining similar tables or figures where this is possible without creating undue complexity.

You are now in danger of getting into the kind of detail that belongs to the final stages of preparing a manuscript for submission, but make no apology for that. It all adds to the desirable impression that the students' results are of paramount importance; also, that tables and figures themselves communicate — communication is not all done with words. Obvious as it may seem, this idea comes to some as strange and new.

Summary of Chapter 4

Designing tables and figures

- Have students finalize the design and content of their tables and figures, checking them against the sequenced section files and making each table or figure a complete unit of communication (Step 11).

- Teach good design of tables and figures, complete with all the surrounding wording, with the aid of a list of do's and don'ts.

5
Topic and sentence outlines

OBJECTIVES:
To discuss the differences between and value of topic and sentence outlines, but without insisting on the use of both.

To point out how a topic outline emerges naturally from the sequencing of items in the ragbags (Chapter 3).

To discuss the use of headings and subheadings.

ASSIGNMENTS:
Make a topic outline from the sequenced section files, and/or make a topic outline from a published paper.

Attempt a sentence outline based on the topic outline.

TIMING:
You will not need more than half an hour to teach the content of this chapter, but there will be plenty to occupy you and the class in discussing the assignments from the previous session.

Here at last we get to putting words down in some sort of order: not quite writing the text yet, but we're moving up on it. Students will have in hand their main headings and possibly subheadings, they will have been developing and refining their ideas through stocking their section files and sequencing them, they will have finished designing the presentation of their data, and they will probably no longer be reluctant writers but writers absolutely champing at the bit to write. There is still one more step that must be undertaken if the writing is to go smoothly.

Step 12 Make a topic outline and consider writing a sentence outline

Topic outlines
The topic outline defines the order in which subjects will be discussed. Much of the order will emerge naturally as the student considers and re-orders the contents of section files, the ragbags. Such re-ordering is so much easier on the computer screen than in the old days of insertions and deletions on paper that the topic outline may be ready almost before Step 12

is begun. However, the point of doing it is to make certain that *all* the requisite thoughts and lines of reasoning are in place, and that no massive rewriting of whole sections will be required at a later stage because a yawning gap has become apparent when the first draft is almost complete.

The topic outline, then, consists of a list of nouns and phrases each of which could serve as a sub-subheading if such things were desirable. Each topic in this list will be the subject of one or a few paragraphs in the eventual paper. Remind your students that the main reason to start a new paragraph is to signal the appearance of a new thought or divergent line of reasoning.

Some experienced writers maintain that topic outlines are confining, and constrain their writing. That's exactly what scientific writing should be: constrained within logical bounds. Exasperating indeed are the papers whose text, especially in the Discussion, marches off in one direction and then makes several right-angled turns before the end of the first paragraph, each naively signaled by the word "However" at the beginning of every other sentence. If the author carefully prepares a topic outline that makes logical sense, and sticks to it, the need for the introductory "However" should shrink to a few well-chosen usages per paper or none at all.

Were your students not scientists, it might be necessary to enter into a discussion of what constitutes logical order, but for most of them their training will make this unnecessary. If any nevertheless remain uncertain, you might suggest using two or even three levels of subheading within the main headings, then first examining all the level 1 headings as a group to see whether they are matched in type and weight; considering all the level 2 headings in relation both to the corresponding level 1 heading and to one another; and proceeding similarly with the level 3 and any further headings.

Sentence outlines

Opinion is divided on whether, before one embarks on the first draft, it is useful to elaborate the topic outline into a sentence outline. What is the difference between them? As we have said, the topic outline is a list of nouns or noun phrases — the subjects, or points, to be successively discussed — whereas the sentence outline is a series of complete sentences stating what the author will actually say about each topic. For some, this comes perilously close to writing the complete draft, hence the hesitation about recommending it. But students who learn to confine themselves to one sentence per topic will be exerting a healthy discipline over their sprawling thoughts, and this should have a lasting beneficial effect on their writing generally.

Ask those students who elect to write a sentence outline to try to compose each sentence in such a way that it can be used intact in the eventual paragraph. Such a *key sentence* (also known as a *topic sentence*) sums up the main thought of that paragraph. It often comes at the beginning of the paragraph, to herald the forthcoming content, or at the end, to summarize (or signal the endpoint of) what has gone before. Occasionally it can even come in the middle, with the words before it swelling to a climax, as it were, and then dying away to a comforting, relaxed conclusion.

Now we are trembling on the verge of a discussion of style, and we don't want to get into those realms until Chapter 7. The discipline of compressing what one has to say into a sentence, and that not an over-long one, is something some students may enjoy, and some may even welcome as "good medicine"; in any case, it is worth discussing the sentence outline as a potentially valuable option.

A good test of the sentence outline of the major assignment, and the logic that underlies it, is to show it to another person, along with the title of the piece, and see whether together they make sense. By this stage, they should.

> ASSIGNMENT
> Make a topic outline out of the re-ordered section files, checking that all topics that need to be included are accorded a logical place in the outline. If necessary, move items from the originally assigned section to another (for instance, from Discussion to Results or from Introduction to Discussion) in order to make the ideas flow logically.
>
> AND/OR (especially for the less experienced):
> Make a topic outline from a published paper.
> (This gives practice in analytical reading, which will be useful when students review their own writing critically later.)
>
> ALSO:
> Attempt a sentence outline, (a) making each sentence encapsulate one subject in the topic outline and (b) making each suitable as a key sentence. Assess the value of doing (a) and (b).

Summary of Chapter 5

Topic and sentence outlines

- Point out how topic outlines emerge from marshaling the notes in section files.
- Show how a topic outline (Step 12), namely a list of nouns and phrases, monitors and controls the logic of the article's structure.
- Point out the value of headings and subheadings on different levels at this stage, even if they are not used in the final article.
- Explain how sentence outlines help some writers, as a further check on logical construction and an exercise in conciseness and precision of thought.
- If students attempt a sentence outline, guide them into making key sentences for eventual paragraphs.

6
First draft and first revision

OBJECTIVES:
To get students to write the first draft in one swing (to achieve unity of purpose), deploying a few given principles for each major section.

To point out that the first revision must begin with a search for flaws in logical progression and misinterpretation of others' results.

To help students avoid getting bogged down in matters of writing style.

YOU WILL NEED:
Copies of a good Introduction from current literature.

Possibly, copies of the Materials and Methods section from a dissertation in the field.

ASSIGNMENTS:
Analyze published Introductions.

Condense the Methods section of a dissertation into journal article length and form.

Write the first draft of the projected article and review it for logical construction.

TIMING:
One session for the teaching element; the length of the last of the three assignments may necessitate delaying the next session and chapter for a week or longer.

Step 13 Write the first draft continuously, collecting references as you go

With topic or sentence outline in hand, or readily accessible on a word processor, the author can now happily proceed to write the first draft. The object of all the careful preparation to date has been to enable the writer to produce the paper as a unified whole, not a series of attempts at what are regarded as the easier passages (Methods, and Results without interpretation) cobbled together like a patchwork quilt. If readers are to grasp the article's message easily and with pleasure, the lines of evidence and reasoning must arc directly into their brains. This will certainly happen if, but only if, the author knows exactly where the article is aiming at any given moment.

The vital piece of advice to your students is: write swiftly, without bothering about graceful phraseology or even choice of words, but continuously, keeping the main lines of thought as laid out in the outline(s) well in sight. The first draft should be completed, if possible, in a single session. This cannot be achieved in a couple of hours after supper. Students must, like every author however experienced, set aside the best part of a day, secluded in a study at home, in a hushed library, or at worst in an office with the door locked and the telephone disconnected.

One thing alone may be allowed to break the continuity. Whenever reference is made to published work in the text — and this can be in highly abbreviated form, e.g. "(Samilowitz 1988)" — a corresponding note must be added to a list in the section file headed References. We shall return to this list immediately after the first draft is finished (Step 14, p 43).

Even if the target journal uses the number system for the references (as most, but not all, scientific journals do), all references in the text of the first draft, and indeed of every draft up until the very last one, should be given in the so-called Harvard form, as in "(Samilowitz 1988)" above, or "(Smith and Jones 1995)". This is both more informative for the co-authors and pre-submission readers as they read and criticize a later draft, and much less hazardous for the writer: there will inevitably come a time when references will be added, deleted, or cited in a different order, and numbers used in a first draft would then have to be changed. If this happens more than once, chaos tends to ensue. (Don't encourage your students to rely on the much-vaunted automatic footnote-processing programs, which can cause an astounding amount of confusion.)

There follow a few principles to observe in writing the different sections.

The Introduction

The key piece of advice here is: *keep it short.* A journal article, unlike a dissertation or a review article, need not survey the whole history of the field, or even a relevant small corner of it, before plunging into new material. Readers know that new articles, however good, are going to add only a tiny crumb to the mass of knowledge already in the public domain — they do not need to be crushingly reminded of the fact by a stately march-past of a vast army of citations.

Disabuse your students' minds of the notion that they must

demonstrate their diligence in searching prior literature by citing dozens of references before they begin. No, the skillful and intelligent approach is to *select* from the mass of previous work what is truly relevant to the new work to be described, making clear just why the investigations were begun.

Good introductions often come in three parts, whether single sentences or short paragraphs. The first delineates the general field in which the work falls, and indicates its importance to pure science or human well-being. The second presents, in main lines only, the findings of others that will be challenged or developed. The third brings out the novel element(s) in the work to be described: the availability of a new technique, say, or an approach to the question from a different viewpoint or with a different population, species, or material. This third paragraph should make explicit the specific question(s) the authors have asked, and point out the advance that has been made.

Above all, the writer must aim to be crisp, not exhaustive, and to entice the future browser into reading on.

If you can find an Introduction that fulfills these aims in a current journal, distribute a copy to your students and have them analyze its good points and any weak ones. Demonstrate, if possible, that the best Introductions are intelligible to readers who are not specialists in the field.

ASSIGNMENT
Analyze one or more Introductions to published articles, either in your own field or tangential to it, listing their good and weak features.

Materials and Methods section(s)
Depending on the complexity of the experimental design and the novelty of the materials or methods used, these two aspects — together with any ethical considerations — may be described in one section or separated into two. The main difficulty in this part of the paper is to decide how much information to provide. Here is where an awareness of the future reader comes in, and where the purpose, scope, and type of the target journal are of particular importance. The best Methods sections describe study design and analysis, not simply the number of rats and the concentrations of chemicals, and the most thoughtful Instructions to Authors actually require such clarification; so recommend that your students think along these lines from the start.

In highly specialized work destined for a similarly specialized journal, less detail is necessary than in a more general, broad-scope journal. Although there is an acknowledged obligation to write this section in such a way as to enable others to repeat or critically appraise the work described, students must take care not to turn it into a laboratory manual.

Almost certainly, the least experienced scientists will be inclined to provide more detail, both in this section and the preceding Introduction, than is appropriate — especially if they have recently completed a dissertation. Help them conquer this inclination by means of the following assignment, for which you will need to choose and copy the Materials and Methods section from a suitable dissertation.

> ASSIGNMENT
> Condense the Materials and Methods section(s) from a dissertation (provided by the instructor) into a form suitable for an article in a specified journal.

This assignment not only gives students practice in extracting essentials from a mass of detail, but brings home the difference between the requirements of the two forms of publication.

Remind your students that the Methods section must contain an explicit statement of any approvals given by ethics committees to embark on the research described.

Results section

By now the results and their significance will have been so thoroughly digested in earlier steps that the section should almost write itself. Emphasize again that the text must not repeat the data shown in the tables and figures, since these will have been constructed in such a way as to tell their own story. The text should make clear the object and purpose of each set of experiments or observations, point out salient features such as an unexpected nonlinearity or the contrasting magnitudes of values obtained, and provide any interpretation needed for an understanding of subsequent sets of results.

Your students may ask whether it is forbidden for the Results section to contain any conclusions or, on the contrary, whether there should be a combined Results and Discussion section. It is impossible to be dogmatic, but I would say that the two sections should usually be kept apart. The Results section should indicate the trend of the author's reasoning without entering into the wider significance of the conclusions, which properly belongs in the Discussion. But authors who stick rigidly to reporting their observations, without any hint of interpretation, in the Results section run the risk of having to repeat the results as they interpret them later. And repetitiveness is anathema to editors and readers alike.

The Discussion

Inexperienced authors really let themselves go in the Discussion and bore the pants off their readers by examining every possible interpretation of every aspect of their investigations. They then emit howls of protest when the journal editor asks that the Discussion be cut by 50%. Tell your students to keep their sights trained on the main object: to lead readers by the shortest route to the heart of the argument. They must not be distracted on the way by useless digressions or the piling of hypothesis upon hypothesis until readers cry, like Macduff, "Hold, enough!"

Thus, the writer should not begin the Discussion by repeating the background already dealt with in the Introduction, but should get straight to the point. If the Introduction announced that the research would ask a specific question or posit a specific hypothesis, the answer or conclusion should be given straight away. Main conclusions should be discussed before any secondary findings; they should, naturally, be related to others' published findings, but not detail by detail.

One's own results are of course endlessly fascinating, and the future is bright with promise as one contemplates them — but usually, only for oneself. It is tempting too to speculate into the wide blue yonder on how one's findings will change the world, but students must be made to recognize that this kind of speculation is for themselves or their intimate friends only, not for public consumption. More to the point is to look the limitations of the work in the face and think about what future research might compensate. Try this as a piece of blunt advice relevant to the Discussion section: *Watch out for symptoms of megalomania.*

Controversial issues should be discussed lucidly and fairly. Where students' results differ from those of others, an explanation rather than a refutation should be sought. Anomalous results for which no explanation is readily available should be clearly pointed out, and internal inconsistencies or gaps in the argument emphasized rather than concealed. Most interesting and valuable to science are the results that open up new possibilities of exploration, and these should be brought to the fore. Some degree of speculation is of course in order in the Discussion section, but it must be reasonable, firmly founded in observation, and subject to test if it is to get past a responsible editorial board.

Your budding authors may wish to cite in the Discussion (or in the Introduction) another worker's unpublished results or observations, as a "personal communication", to bolster the argument or justify an interpretation. Stress that the author must get permission from that person, in writing, to do so and must show him or her how the quotation has been used, in context. The author should start asking for the permission very soon after completing the first draft, along with the permissions to acknowledge personal services in the relevant section. Should it be desirable to reproduce a *published* illustration in support of the arguments, permission must of course be obtained not only from the original author(s) but also from the copyright holder (e.g. the publisher) if different.

Constructing the list of references in complete form

When the first draft is finished, students should banish it from sight and return to their notes of the references they wish to cite. They must then make sure that they possess a photocopy of each article cited.

The purpose of doing this is twofold. First, some students tend to think that "tidying up the reference list" is a mechanical, librarian-type job that can be left to the end or to obliging library assistants. They are then in trouble when, a day before they are due to go on holiday after submitting the manuscript to a journal, they discover that the relevant volume is missing from the library — stolen, lost, or at the binder's. This calamity is averted if a photocopy is safely in their hands.

Secondly, and more profoundly, it is essential at this stage to re-read the cited references *in their entirety*. Often, the key article that prompted the entire investigation has been read and internalized many months, perhaps years, before. Unfortunately, memory plays tricks, especially on material that has been re-worked in one's thoughts over time, and one can find oneself quoting the exact opposite, or a twisted version, of what that article originally asserted. Subtle or gross variants may have crept into the methodology adopted from others, and significant disagreement with the original results may become apparent from the re-reading. So although it seems harsh to insist on this re-reading, assure your students that it is time enormously well spent, especially given the probability that authors of the papers cited will be among the reviewers of the submitted article.

> ASSIGNMENT
> Write the first draft of your major assignment swiftly, eliminating distracting activity, and bearing in mind the guidance you have been given.

This assignment is, obviously, of major proportions, but you will not be asking to see it at this stage, and you can move on to give your students a sneak preview, as it were, of what is then going to happen to it.

Step 14 See if the first draft needs major alterations

The period of re-reading of key references and reflection on their significance also ensures that time will elapse between the writing of the first draft and its first revision — a period in which it lies dormant in a drawer or deep in the computer, but undoubtedly seething in the subconscious. Every writer in the world, save daily-press journalists (who don't have the option), insists on the importance of this dormant period.

When students return to the first draft, having written off for the relevant permissions to quote and having done all the re-reading of references and the re-thinking so engendered, they should not be dismayed by the roughness of diction in the first draft or the bumpiness of the sentences: what they must diligently seek out in the first instance are flaws in the logic, faults in the flow of ideas, and misrepresentation.

Logical flaws

If the student's outlines were carefully constructed and diligently followed, there should be no problems of logic with their main lines of reasoning. Now is the moment to look at subsidiary arguments and to decide whether, after all, there is scope for minor adjustment; also, whether individual sentences, however roughly expressed, make logical sense both within themselves and in relation to their neighbors.

Think not only of logic but also of relevance: tempting lines of argument that do not contribute to the article's purpose should be ruthlessly culled.

Misquotations

Authors should inspect with particular care any statements made about the work of others. Here is where the re-reading of all the cited papers again comes in, for woe betide the author who misquotes or seriously misinterprets published work, which will be familiar to at least some readers, including reviewers.

Your students will doubtless protest that there is little or no chance of their sinning in this way, for they have been living with those papers in their minds for months. That's just the trouble: while someone else's results are in one's mind they tend to swirl around into shapes that fit one's prejudices or favored hypotheses. The main source of misrepresentation is faulty memory, not cunning or guile. Persuade your students to re-read all the papers they have cited, betting them $100 that they will as a result either change what they have said about them or delete one or more of them as irrelevant. I'm sure you'll come out ahead. And science will be the better for it.

Summary of Chapter 6

Writing and revising the first draft

- Persuade students to write the first draft continuously from beginning to end, with their outlines before them, deferring language problems till later.

- Insist that they note each citation in a separate file as it occurs, and at the earliest opportunity check that they possess a copy of the complete article to which each refers.

- Give the following tips in composing the first draft:
  Introduction
  - Keep it short and intriguing.
  - Compose it in three parts (present knowledge, new approach, hypothesis or question posed).

  Materials and Methods
  - Include the appropriate amount of detail, being aware of the readership of the target journal.
  - Stick to the logical order worked out in the topic outline as suitable for the intended audience.
  - Be sure to specify what permissions regarding the ethics of procedures were obtained, if relevant.

  Results
  - Allow the data to speak for themselves.
  - Do not rigidly save deductions for the Discussion.

  Discussion
  - Watch out for symptoms of megalomania.

- Advise students to put the completed first draft away while they attentively re-read all references cited.

- Have them then thoroughly review the draft, looking in particular for flaws in logical progression and for faulty quotation or misremembering of others' work.

7
Editing the text and rewriting title and abstract

OBJECTIVES:
To help students get illustrations and their wording into final form.

To enable authors to view their text dispassionately by editing it for clarity, precision, and conciseness using four style guidelines.

To inculcate this method of reading scientific texts critically as a tool for long-term future use.

YOU WILL NEED:
Council of Biology Editors Scientific Illustration Committee. *Illustrating science: standards for publication.* Bethesda (MD): Council of Biology Editors, Inc; 1988. Especially Chapter IV, which gives different types of line drawings to illustrate numerical values.

YOU SHOULD READ:
William Strunk Jr and E B White. *The elements of style.* New York: Macmillan; 1979 and any subsequent editions, especially Usage Rules 10, 12, and 13.

YOU SHOULD CONSULT:
Mimi Zeiger. *Essentials of writing biomedical research papers.* New York: McGraw-Hill; 1991.
This provides a wealth of examples and exercises to improve writing style.

ASSIGNMENTS:
Apply four style guidelines to gain practice in editing your own writing.

TIMING:
At least two sessions, and considerable time set aside to consider revised first drafts outside class.

When students are completely satisfied with the logical structure of the first draft, encourage them to revise their text into good English before they send it for informal review by colleagues. You can help your students revise for precision and clarity by teaching them the editing techniques described in Step 16, which enable them to act as though they were working objectively on someone else's words. I suggest four editorial *style guidelines* that have stood the test of time. While students master these techniques, they should be setting any available professional technical draftsmen to work on preparing their illustrations.

Step 15 Have the illustrations prepared in final form for the target journal

Graphs and photographs need to be prepared to professional standards if they are to be adequately reproduced in a printed journal or viewed on the Internet. If the student's institution has an illustration department the raw material should be sent there without delay, unless there is any likelihood that co-authors will want changes, in which case this step should be held over until after Step 18. Any special instructions from the target journal should of course be sent with the order.

Computer-generated graphs with crude lettering on the axes and jerky "curves" joining points marked by unreadable symbols are not acceptable for journal publication and should be re-drawn professionally.

Micrographs should be marked for cropping so that the field of interest is centered and the image is of a size and shape suitable for reproduction in the target journal. The degree of magnification will need to be stated in the legend and checked at proof stage in case the figure has been reduced by the journal; this is easily done if a magnification bar, identified in the legend as representing 10 μm (or whatever), is included (see Step 11).

Halftones can nowadays, with computer scanning, be accurately printed more readily than in the past, and color printing is less expensive than previously. Nevertheless, both processes can be costly, and authors should be aware that journals may charge for extensive use of either process.

Tables, if complex, may best be supplied to the journal for photographic reproduction like graphs. If authors want to do this they must first ask whether the journal accepts this procedure and if so, study the target journal's conventions as to:

- what sort of typeface is used for the title and various headings within the table,

- where in the table units of measure should be placed,
- how thick any horizontal lines should be, and
- whether vertical lines between columns are used,

and instruct their illustrator accordingly; alternatively, they should construct the table using their own computer software and have the result photographed and supplied as a glossy print.

At this stage a decision needs to be made on whether to include all the data or to drop minor facts that neither support nor contradict the main conclusions. Advice to authors in the past has always been that extensive collections of data, which the author feels may be wanted in the future but which the editor does not want in the present, may be deposited at a central archive. This has become extremely easy with the advent of the Internet, though there are certain formalities to be observed before one obtains permission to make the deposit.

However, students should not run away with the idea that publication in an electronic journal, in which a lengthy article can be published as inexpensively as a short one, exonerates them from the duty of making their tables and figures few and comprehensible. A sure way to deter readers from reading an article is to flood them with data.

Step 16 Polish the prose

Only now do we focus on wording. The first draft will have been written not exactly in a rush, but with a purposeful drive to get the concepts, the facts, and the arguments set out in a logical order, without seeking the most illuminating way of expressing them or finding exactly the right words. Tell your students that the most efficient way of straightening out the kinks and aiming their message into each reader's brain without fear of misunderstanding is to edit their first draft systematically, deploying four guidelines aimed at sound scientific style.

I derived these guidelines through an analysis of the turgid and obscure prose I saw as an editor, in papers submitted by scientists whose brains I knew were first-rate but whose style was awful. It had been corrupted, I believe, by a stodgy mode of writing that became conventional for factual writing during the first half of the 20th century.

The style guidelines (SG1 to SG4) are:

SG1 Write simply and concisely.

SG2 Check the meaning and relationships of every word.

SG3 Seek and release hidden verbs of action.

SG4 Split up noun clusters and stacked modifiers.

Style Guidelines 1 and 2 apply to any kind of writing; 3 and 4 attack faults particularly common in scientific writing. I have called them guidelines in this book, rather than rules as in the 1960s, because although "rule" is brisker it is also more directive than was ever intended, and prose can actually sometimes be clearer and more effective if a so-called "rule" is broken.

SG1 Write simply and concisely

Inexperienced writers in science often aim at an elevated style which, they think, suits the formality of publishing for posterity. They seem bewitched by orotund polysyllables and employ cumbersomely coy phrases that are heavily dependent on verbs in the passive voice, perhaps because somewhere deep down they consider short words childish and active verbs shocking. They have forgotten that the object of publication is to communicate to a wide-awake reader, whereas the kind of language they favor is soporific. Try to get them to use lively, vibrant words and phrases.

You can't give better advice than Strunk's. William Strunk Jr was a professor of English at Cornell University in the first half of the 20th century, who was rescued from oblivion when one of his former students, the writer E B White, republished in 1959 the short book Strunk had had privately printed in 1918. "This book", says the preface, "aims to give in brief space the principal requirements of plain English style."

Strunk's most inspiring rule, as far as I am concerned, is

> **WRITE WITH VIGOR.**

And "vigor" is the watchword your students should always carry in the back of their minds as they tackle the halfhearted, labored phrases their first draft is doubtless awash with.

Apart from its power to communicate, vigorous writing precludes ambiguity, which can be the very devil for scientists who wish to understand exactly what was done and perhaps to replicate it. Give a few examples to illustrate how using this guideline can abolish ambiguity, then set other examples for simplification as a minor assignment. Thus:

> Insufficient data are at present available to completely negate the possibility that removal of the above-mentioned substances from the circulation is not a factor of importance

becomes, after much anxious thought, something like:

> Removing these substances from the circulation might affect the outcome; we cannot yet say.

Likewise,

> In chromatography studies pertaining to the identification of phenolic derivatives, drying of the paper gives less sensitive visualization

can only mean

Phenolic derivatives can be seen more easily if the chromatography paper is left wet

but what a lot of effort goes into working that out!

Now we encounter a roundabout construction that has become almost obligatory, I can't think why, in scientific papers. It's as though "than" is a forbidden, four-letter word:

The blood pressure of the experimental rats is *increased compared to* that of the controls

meaning

The blood pressure of the experimental rats is *higher than* that in controls

or, even simpler,

The blood pressure is higher in the experimental rats *than* in controls.

Advise your students to regard the phrase *compared to* as the first of many "warning words" or phrases (see Appendix 3, p 172) that should act as a red flag urgently signaling the need for corrective action. Here is another one, *with respect to*:

Serum cholesterol was elevated *with respect to* the Taiwanese population

meaning, simply:

Serum cholesterol was *higher than* in the Taiwanese subjects.

Now see how the use of *compared to* can lead to ambiguity:

Enzymatic activity is decreased after drug treatment compared to those on placebo.

Does this mean

Enzymatic activity is *decreased more* by the *drug* than by placebo treatment

or

The drug, but not placebo, lowered enzymatic activity

or just the following?

Enzymatic activity was *lower* in patients given the drug *than* in those on placebo.

Students must learn that the verbs *increase, raise, elevate,* and their antonyms contain a time element, and actually mean that the change in value takes place over time or relative to a previous state, for instance before the corn harvest or prior to medication. The words are too often used, however, when two measurements were made at the same point in time, and all that can be scientifically stated is that something was higher or lower than something else.

Here is an even more convoluted, inaccurate statement along similar lines:

> In comparison to the aortic endothelial lesions, the size of the coronary artery lesions was smaller

with the added absurdity of a size being smaller, rather than the lesions. Only clothes and shoes come in smaller sizes. How do we correct this? The most straightforward translation would be

> The aortic endothelial lesions were smaller than the coronary artery lesions

but this, though simple, is not concise. Better would be:

> Endothelial lesions in the aorta were smaller than in the coronary arteries.

ASSIGNMENT ON STYLE GUIDELINE 1
See Appendix 1, p 131.

Students should be aware that their first attempt to write simply may result in a passage that is far from concise, indeed is verbose. But urge your students to attempt simplicity first, for achieving it is the first step on the road to complete understanding. Then, have them convert this kind of simple, grammatically correct, but tedious passage into equally simple but more acceptable form:

> The numbers of enucleated cells in vaccinated and nonvaccinated mice were determined both at four and eight days after inoculation and (or) the beginning of fasting. The number of enucleated cells in vaccinated mice was seen to be greater than in nonvaccinated mice and to increase from four to eight days after inoculation, whereas the number of such cells examined under the same conditions in nonvaccinated mice actually decreased during the first part of the experimental period and then increased from four to eight days, but not to the same extent as they did in mice that were both vaccinated and fasted. Consideration of the numbers of enucleated cells in all four groups, see Table XX, reveals that the effect of fasting seems to have been superimposed upon the effect of prior vaccination, at least in the second portion of the experimental period.

The brightest students will, I hope, tell you that most of this should have been obvious from the imaginary Table XX, and that the only prose needed is contained in the last sentence, duly shortened.

You may meet some resistance to your crusade for simplicity, as opposed to pomposity, with the accusation that you are advocating the abandonment of polysyllabic technical terms. Of course, you are not: it is the unnecessarily polysyllabic *non*-technical terms you are attacking. Technical terms, properly understood and used, can be among the most concise in the language, because they pack much meaning into few letters. But beware of using technical terms that are only half understood: as Herman Melville says, "A man of true science uses but few hard words, and those only when none other will answer his purpose; whereas the smatterer in science thinks that by mouthing hard words he proves that he understands hard things." Surely no one in your class wants to be accused of being a scientific smatterer.

Probably your students will need no more convincing. But for the continuing doubters, or for those who have become interested and are realizing for the first time the multitude of pitfalls before them, recommend the following reading:

> Sir Ernest Gowers. *The complete plain words* (1954). Harmondsworth: Penguin Books; 1987. Especially the entries *case, position, situation, conditions, level.*
>
> Sir Arthur Quiller-Couch. *The art of writing* (1916). New York: Capricorn Books; 1961. Chapter V "On Jargon".

Even if these books are out of print in your locality, most good academic libraries will still carry them.

This small amount of reading in the great classics on writing style will do no one any harm. Never mind that the style of these books is redolent of an earlier, calmer, and more gracious age: a little breathing space in our hectic, active lives may actually promote our ultimate aims!

Some faults of pretentiousness can also be attacked in the following style guideline.

SG2 Check the meaning and relationships of every word

This guideline is, admittedly, something of a catch-all, and examining the context of every word in the first draft can be time-consuming. But if one applies the guideline first to verbs, a whole group of common errors can be systematically eradicated. Thus, asking every time "What is the subject of this verb?" brings to light the following kind of non-agreement:

> Lowering of glucose levels remain the primary goal of therapy

and absurdities of juxtaposition such as:

> This value was found by Smith in rabbits who reported that . . .

as well as the well-nigh ubiquitous dangling participle:

> After removing the intron, the protein was mainly localized within the plant nucleoplasm.

Based on these calculations, we concluded . . .

The protein has been expressed at high concentration in tobacco plants using cytoplasmic potato virus.

Now, you may find your students at sea when you introduce the term "dangling participle" (also "hanging", "floating", or — best of all — "unattached" participle). Ask them whether you need to explain it, and, if you do, use Appendix 4 to review their knowledge of grammatical terms respecting verbs, particularly the active and passive voice, and stress that every verb must have a subject, implied or explicit. They should then have no trouble correcting the above examples or detecting another type of "warning word": words that end in *ing*. The present participle is the one most often misused, though the past participle *based on* (instead of *on the basis of*, as in the example above) is common too.

Another common fault, namely omission of the auxiliary verb *was* or *were* in a series of actions involving different subjects, is also caught by SG2:

The rats were killed and their blood pooled.

Here, the auxiliary *was* must be inserted since the implied *were* is plural while *blood* is singular. The safest rule is to insert the auxiliary whenever the subject changes, even if the new subject matches the first one in number, thus not

The rats were killed and their hearts measured

but

The rats were killed and their hearts were measured

because in the first of these two sentences the reader is subconsciously expecting the hearts to measure 2 or 3 cm instead of *being* measured.

The following kind of mistake may not be so easily discerned by an author who has written something like

The adsorption is completed in 15 min. This greatly reduces previously reported adsorption times.

But again, if he or she asks "What is the subject of this verb?" (here, *reduces*), that author will quickly perceive the idiocy of implying that a fact in the past can be changed by a finding in the present.

You may introduce some other pet peeves at this point. Mine include the use of *varying* (which means changing with time) when *different, various*, or *varied* is meant; *due to* instead of *because of*; and *following* or *post* when *after* is the correct English preposition.

ASSIGNMENT ON STYLE GUIDELINE 2
See Appendix 1, p 132.

SG3 Seek and release hidden verbs of action
Much scientific writing, even if precise and commendably unpretentious, is lifeless. A prime reason is that scientists, overwhelmed by the desire to appear objective and impersonal, succumb to the allure of abstract nouns in place of verbs in the active voice. Instead of writing "we poured the liquid" they shelter behind the passive voice and write "pouring was effected". And when the action has to be further qualified, they get into the most tremendous tangles: instead of "we poured the liquid carefully down the drain" they mumble "pouring of the liquid down the drain was effected using the utmost care", throwing in the dangling participle *using* for good measure.

All writers on style advocate writing with vigor. H W Fowler in his classic *Modern English Usage* (Oxford University Press, first published 1926) does; so does Strunk; F L Lucas, in a wonderful book now out of print entitled simply *Style* (1955), says "It is not much use making your reader see, if you also make him yawn". Other popular advocates of vigorous writing are Jacques Barzun, in *Simple and Direct: A Rhetoric for Writers,* and William Zinsser, in *On Writing Well.*

How does one put vigor into one's writing? By using strong, active verbs. Hiding them away in abstract nouns (*separation, precipitation, extraction, centrifugation, expression*) drains them of their force. Warn your students that the very construction of topic outlines, which consist of a series of (usually abstract) nouns, tends to draw them into this fault.

But it is easily rectified. In this part of the self-editing step one can go on a "seek and destroy" mission of a large proportion of the abstract nouns derived from verbs, emitting the battle cry "Release the hidden verb!" I say "large proportion", because it would be absurd to attempt to dispense with abstract nouns altogether. The warning flags here are some tell-tale colorless words that always suggest there is a hidden, active verb nearby (for instance in *separation was effected,* better expressed as *the isomers were separated*):

accomplished	effected	made
achieved	experienced	obtained
attained	facilitated	occurred
brought about	given	performed
carried out	implemented	proceeded
conducted	indicated	produced
done	involved	required

Spotting abstract nouns and transforming them into clauses based on the hidden verbs will do wonders for the vigor and liveliness of the writing, and often simultaneously reduce pomposity and increase simplicity. Practice at doing this soon makes perfect.

ASSIGNMENT ON STYLE GUIDELINE 3
See Appendix 1, p 133.

When students have become adept at recognizing active verbs hidden in abstract nouns they will take delight in identifying other roundabout phrases based on verb forms: they will replace *is explanatory of* by *explains*, and *is the resultant of* will become *results from*. Substituting the active verb form is almost always desirable, and will generally produce a simpler and more succinct text. Furthermore, using the verb in the active voice obliges the writer to specify the subject of the verb, which the passive voice leaves obscure.

So what is your advice about using the first person in scientific papers? Since most papers now have more than one author, the first person *singular* rarely comes into question. What about the plural? In the Methods section, actions taken by technicians or others who are not in the byline should not be given the subject *we*, a pronoun that must always be taken to imply the paper's authors. In this case, the passive voice may actually be appropriate, but students will find on close examination that the actual subject of the verb is rarely the human beings taking the action but the inanimate or living objects acted upon. So they should still concentrate on releasing the hidden verbs (and activating passive ones) wherever they can. If after doing so they find that the earlier sentence was neater, encourage them to revert to it.

SG4 Split up noun clusters and stacked modifiers

In English, nouns can be used adjectivally, as in *cat flap*, meaning a flap for cats, or *ionization chamber*, a chamber in which ions are produced. If a further noun is added to qualify the first adjective, as in *lead ionization chamber*, we get a noun cluster, in which it is no longer clear whether lead is being ionized in the chamber or the chamber is made of lead; as further nouns are added, the cluster gets increasingly unintelligible, as it is less and less clear which noun is qualifying which other one. For some inscrutable reason, noun clusters and stacked modifiers beguile many scientists as irresistibly as Lorelei did the sailors on the Rhine. Take this one: *Radioactive surface contamination monitoring workshops*. Easy, a skeptic may say, to sort it out with a few prepositions: *Workshops on the monitoring of surface contamination by radioactivity* — but why should the readers be condemned to wait for the last word before they figure out what you're talking about? Curiously, unraveling these concatenations often starts by putting the last word first.

It is worth while introducing another technical term here, the "compound adjective", as in *hot-wire detection system* (a system for detecting gases emerging from a gas-chromatographic column), which is too often deprived of the hyphen it needs to link the related words together. Unfortunately, some journals and publishers have adopted, without consultation, the totally arbitrary policy of deleting all hyphens, quite unnecessarily exacerbating the problem readers face in squeezing the meaning out of strings of words an author has flung before them.

Here are some real (not invented) examples. Can your students tell which words relate to which others?

heavy beef heart mitochondria protein
a radium containing argon ionization chamber
undetectable (<50 nmol/l) midnight plasma cortisol concentration
skin water accumulation
skin inflammation reactions
mouse keratocyte cell line
preformed active pre-IL-1
to maximize TBT-induced IL-1 production
goat anti rabbit IgG, alkaline phosphatase conjugated

The habit of stacking modifiers and eliminating hyphens leads to a common kind of phrase that conveys the exact opposite of the intended meaning:

methane containing mixture
tumor bearing mouse

These phrases would become acceptable with the judicious insertion of a hyphen, but simple inversion of the word order often suffices to restore logic and sense. In grosser cases, such as

silica gel coated glass fibre paper chromatography

the only solution is to split up the stack, move the noun now at the end to the front, and deftly explain the relationships by inserting prepositions:

chromatography on glass fibre paper coated with silica gel.

Although this phrase is longer than the original, thus seemingly violating SG1, it is now unambiguously and instantly intelligible.

ASSIGNMENT ON STYLE GUIDELINE 4
See Appendix 1, p 134.

Students may have difficulty with this assignment because one needs to get inside the author's head if one is to be completely sure that the translation is right — which neatly illustrates why authors should not write in a fashion that not only slows the readers up but may leave them baffled in the end.

Exercises on the application of all four guidelines

Of course, texts in the real world will show a mixture of stylistic faults, often in the same sentence. At this point you can offer your students an exercise demanding the application of all four guidelines to single sentences containing a medley of faults:

ASSIGNMENT ON ALL FOUR STYLE GUIDELINES
See Appendix 1, p 135.

Or you can let them loose on whole passages, perhaps beginning with the faulty abstracts in Appendix 2 and then moving on to complete papers (also in Appendix 2) in which each element — title, abstract, text, tables, references, over-all structure — is open to criticism. Which you choose will depend on the time available and the degree of enthusiasm evinced by your students. It is important not to distract them from editing their own first drafts, but before encouraging them to do so you must judge whether they have mastered these editing tools and are ready for the task. Otherwise there may be disappointment and disillusion all round.

To assist the students, now and in the future, hand out copies of Appendix 3 (or your own variant of it), which lists the words and phrases to avoid or to notice as warning flags. Of course, do point out that they can be telling enough in their place, especially if sparingly used.

You must be prepared to give a great deal of helpful feedback at this stage. It will be easier for you if you ask the students not only to print out their revised first draft double-spaced and with wide margins for your comments, but to provide the unrevised first draft for your reference should you encounter difficulties in comprehending the revised form!

Step 17 Rewrite the title, and structure the abstract

The many revisions up to this point may have subtly changed the thrust of the article, and the wording of the title must now be reconsidered. Students should review the working title devised at Step 6, and modify it if necessary in the light of the following considerations.

Characteristics of good titles

Titles usually consist of a qualified noun (*Effect of, Decreased susceptibility to, Blunted reaction to*), but one can occasionally use statements (*Glia increases trimethyltin cytotoxicity in hippocampal neurons through release of TNF-a*) or even questions (*Is measurement of antioxidant capacity clinically useful?, Has the case for genetically modified pesticides been proven?*). But caution your students to consult the target journal's Instructions to Authors, because some (e.g. *Journal of Virology, Journal of Bacteriology*) outlaw such titles.

It is important to get as much information into the title as possible while keeping it short and, if possible, arresting. Think of it from the point of view of scientists *outside* the field who may be unfamiliar with the intimate jargon of the subspecialty but who can be enabled, if one makes the title intelligible to a wider audience, to do some lateral thinking and perceive the article's application in their own field.

These somewhat conflicting aims may often be resolved by the judicious use of additional key words or phrases (sometimes known as indexing terms, which is what they are). Clever writers do not simply repeat the words in the title, perhaps varying them slightly, but use the key words to point to unsuspected charms in the body of the paper. Advise students not to exceed the number of key words or phrases permitted in the Instructions to Authors; if they do, those at the end of their list will simply be deleted, so they might as well make their choice now.

Titles are more important than most people realize. They may be the sole means of catching the attention of those who simply scan titles in *Current Contents* to determine what they will look at further. But the title must not mislead, or claim more of interest than the paper contains. It is worth writing down a selection of possible titles, ringing the changes on a number of essential words or concepts, and trying them on colleagues, before settling finally on the one that is inevitable and right.

Rewriting the abstract

Now the working abstract (Step 6) has to be rewritten into an abstract that is no longer for the guidance of the author but for the edification of the wide world. Again, the abstract may be the sole means of luring into the article the reader who found the title interesting enough to turn to the abstract. Remind students not to think in terms of the regular reader of the target journal, who may well scan all the papers in the current issue, but the scanner of up to 100 journals a week, who will spend a few seconds on each title, and perhaps a minute on an abstract before deciding if it is worth bothering to obtain the whole paper.

This is why the main lines of an abstract, particularly the purpose of the investigation and the conclusions drawn, must be bold and clear. To aid readers, and especially those initial readers the editor and reviewers, some journals now require the abstract to be structured, with the material placed under a series of prescribed headings. The headings differ according to the nature of the article, and not all of them have to be used, so the process is not as rigid as it sounds at first, and it does ensure that all the vital information is included.

Obviously, students will have to study the Instructions to Authors to determine whether structuring the abstract is a requirement before they finalize their text. If they are wise, it will have been one of the special requirements they noted at Step 7, and they will have made sure that they have the information that is required under each heading. Even if the journal does not require structured abstracts, there is nothing against submitting a structured abstract if this seems to suit the material. However, insertion of the headings does tend to lengthen the piece, and authors must abide by the word limit. An alternative is to use the headings to build the abstract, and then delete them to make a continuous text.

Whether structured or not, the emphasis in an abstract will be on answering the following questions.

- Why was the work undertaken?
- What new material (species, method, type of chemical) was used?
- What were the results? (in broad terms, not just a string of values with standard deviations)
- What are the conclusions, and why are they noteworthy?

The abstract must not make claims that are not in the paper, nor may it indulge in the kind of speculation and hypothesizing that belong in the text of the Discussion. But it must try to be interesting, and point up what is original about the work.

You may be asked what the difference is between an abstract and a summary. This is easy: an abstract stands alone, comes at the beginning, and must be intelligible without further acquaintance with the article; a summary is sometimes appended to a very long article or to a chapter (as in this book) and summarizes points for the reader who has read what precedes it. Confusingly, a few journals still label the abstract preceding the paper "Summary", but this usage is archaic and should be discontinued. An "executive summary", which comes at the beginning of an informative document or position paper, also meets our definition of an abstract, the essence of which is self-sufficiency.

Although this rewriting of the title and abstract comes at the end of the drafting author's work, it should not be despised or skimped. Bring home to the students the difference between writing the working title and abstract as a guide to oneself, and writing a possibly very different title and abstract for others' consumption. And tell them not to feel guilty or discouraged if the "before" and "after" pieces differ greatly: they are bound to, given the different audiences for whom they are intended.

Summary of Chapter 7

Editing the text and rewriting the title and abstract

- Advise students, before they embark on the editing phase, to make (or have professional draftsmen make) any graphs and halftones in a form suitable for submission, in conformity with the target journal's Instructions to Authors (Step 15).
- Teach them the most objective way of revising one's own writing (Step 16), namely by editing it using Style Guidelines 1–4:
    - SG1 Write simply and concisely.
    - SG2 Check the meaning and relationships of every word.
    - SG3 Seek and release hidden verbs of action.
    - SG4 Split up noun clusters and stacked modifiers.
- Work through exercises illustrating the use of these guidelines.
- Assign some longer pieces of faulty scientific prose (see Appendix 2).
- Hand out Appendix 3 (warning words and phrases).
- When they are ready, have students edit the first draft of their major assignment.
- Describe how to rewrite the title and structure the abstract (Step 17) for public consumption, recognizing their importance in attracting readers.

8
Informal review, final drafts, and submission

OBJECTIVES:
To introduce a method of requesting review by colleagues at three levels of closeness, and to describe how to receive and act on the advice given.

To induce respect for the paramount importance of Instructions to Authors in the target journal.

To provide a checklist of items to be included with the manuscript in submitting it to the journal editor.

YOU SHOULD CONSULT:
X Li and N B Crane. *Electronic styles. A handbook for citing electronic information.* Medford (NJ): Information Today; 1996.

International Standards Organization. *Standard for bibliographic references — electronic documents or parts thereof* (ISO 690-2:1998).

When any author has completed revision of the first draft, he or she should show the resultant second draft to several other readers for comment.

Step 18 Request private review by three independent critics and your co-authors

Three types of person will have useful criticism to offer: a scientist in the same laboratory, a scientist down the hall, and a non-scientist friend.

"Scientist in the same lab" is shorthand for one who is active in the same subspecialty as the writer of the first draft. This first advisor will most probably be the co-author(s) of the article. However, readers in this category could also include a close colleague who is familiar with the pitfalls of this line of investigation or limitations of the methods used, but who has not participated in this investigation.

There is theoretically no limit to the number of people who could be invited to comment from this standpoint, except that the time available is not infinitely extensible. It is important to state a deadline by which the comments are needed — but this can be elastic, especially if readers other than co-authors are being asked to do a favor. Above all, the comments must be in writing. Few authors can avoid becoming defensive in response to oral criticism, and instant rebuttal promotes neither rationality nor good relations.

By "scientist down the hall" I mean the scientist who is in the same general area of science but not in the same (sub)specialty. Such readers are particularly important if the target journal is one of wide scope, which broader-gauge readers may be expected to scan. They should be asked to comment on whether the article is readable and easily intelligible; if not, wherein lies the difficulty; and whether the article contains a suitable degree of detail for non-specialists.

Similar questions should be put to the "non-scientist friend", spouse, or partner. Even if that person finds the science impenetrable, at least he or she should be able to understand the Introduction and get a glimmer of the significance of the work undertaken, as well as commenting on any residual grammatical slips, apparently awkward phraseology, inordinately long sentences or paragraphs, non-parallel constructions, unequal weight of subheadings, and the like.

Students must remember to supply to all these readers the *complete* manuscript, with all the preliminaries (title, authors' names *and affiliations*, key words, and address for correspondence) and attachments, including the complete list of references and especially the tables and figures. It hardly needs stating, after what has gone before in this book on the subject, that the so-called illustrations are actually the heart of the article. To supply the text alone for comment is absurd (though I have seen it happen).

What to do when the comments come back

After all the careful thought and work that has gone into the composition and editing of the first draft, the fact that some of the readers' comments are critical will come as a disagreeable shock. Impress on your students, therefore, that they must thank the critic for taking so much trouble *before* reading the comments, then go away and consider them quietly at leisure. They will have been given in good faith and should be pondered in the same spirit, and preferably slept on before any action is taken.

Step 19 Re-read the Instructions to Authors and make any necessary adjustments

At the same time as making changes in the light of the comments received, the author should take pains to put the manuscript into the form suitable for submission. Your students will have absorbed and acted on the salient features of the Instructions to Authors in the target journal in getting thus far, but they will now need to pay attention to minutiae like the exact form in which to put references (noting that citing electronic information requires extra elements to give the date of access as well as the date of original publication — see the references at the head of this chapter), whether the "right justification" function on their word processor should be switched off, which pages should be numbered and whereabouts on the page, whether the abstract and key words or phrases should be together or on separate sheets, whether each main section should begin on a new page, whether a handwritten note should be made in the margin where each table or figure should be inserted, and so on. Urge your students to study the Instructions to Authors line by line at this point.

Step 20 Revise as many times as necessary

With all of this in mind, the student can embark on the changes asked for by the co-authors and suggested by other readers, at the same time putting the paper into final form. This produces the third draft. This, and any subsequent drafts, should be shown to the co-authors only. Tell your students to brace themselves for the likelihood that there will be a fourth and even a fifth draft before all co-authors are satisfied. Reassure them that there is no shame in such numbers: all the best authors attest to the need for several revisions, even of work that has no co-authors. However, a final draft will eventually be reached, and the wretched thing, or masterpiece (depending on one's temperament), can be submitted.

Step 21 Submit the article to the journal

Some journals have such a detailed set of requirements for submission that they append to their Instructions to Authors a checklist for authors to use. The number of hard copies is specified, together with the type of diskette and software to be supplied. Large international journals often have regional or subject-specialized associate editors to whom the paper should be sent rather than the editor-in-chief. One of the authors must be named on the title page, with his or her address, for correspondence and reprint requests (this may or may not be the same as the author with whom the editor is to correspond, which will be specified in the submission letter). Other types of requirement are discussed below.

Specification of reviewers to be used or avoided

Some journals invite authors to identify reviewers who would be competent to judge the article, or even reviewers the authors want to avoid — never, however, with a commitment on the editor's part to accept the suggestions.

Discuss with your students whether they should submit such names. There is no obligation to do so, but there could be advantage if the authors know that the field is super-specialized or new, so that the editor cannot be expected to have readily to hand a list of reviewers whose knowledge is particularly matched to the subject; or if, on the other hand, the authors know that a rival investigator, who would be an obvious choice as far as knowledge is concerned, is hostile. Even if the editor does use that person, naming him or her in a negative sense will alert the editor to the possibility of prejudice, which the editor can take into account when receiving the comments.

Some scientists are paranoid about the possibility that reviewers will steal their work or delay consideration of the manuscript while they complete results of their own, and publish before the article submitted breaks into print. Try to soothe their fears: although there have been a small number of cases of each of these crimes, the perpetrator quickly becomes known to responsible editors and eventually to others, so the risk becomes one that is hardly worth taking. There are ways, too, of establishing priority, primarily through oral presentation within the authors' institutions or at unpublished conferences, though authors must be careful in the latter case to avoid redundant publication (p 15) if these presentations are unexpectedly published. Complaints should certainly be made to an editor if an author detects publication of data identical with or closely similar to those submitted to an editorial office in the recent past. Editors have the means, and the responsibility, to follow up allegations of this kind and to blacklist a culpable reviewer.

However, the essence of science is openness, and reproducibility. Poor indeed is the scientist who cannot afford to relinquish the credit for pushing knowledge just that little bit forward. If those who have an influence on a scientist's future know the truth about the work that has been done and the strength of the results obtained, even if someone else appears to have got there first, that should be sufficient for the complainant. Assure your students that talent will out in the end. If they but guard their own integrity, their uprightness will ultimately be rewarded.

Signature of all authors

Whether or not the journal provides a checklist, certain actions and documents are always required. For multi-author papers, although one person actually sends the manuscript and undertakes to manage subsequent correspondence and revisions, all the co-authors must sign the submission letter or a dated accompanying sheet agreeing the content of the paper as submitted.

Statement of possible conflict of interest

Many journals nowadays also demand that all authors state what financial or other interest they may have in a drug, chemical, or cell line under investigation or produced in the course of the experiments — or if they might in any other way benefit, directly, from publication of the paper. This almost always relates purely to commercial exploitation — authors are not expected to declare that they hope to get a fellowship, invitation abroad, or coveted appointment as a result, since these are accepted motivators in the academic world! The probability of commercial benefit need not affect the editor's decision adversely; what *will* do so is the failure to state it. Likewise, authors will not impugn their integrity if they openly acknowledge financial support received from a commercial organization or trade association to carry out research that could increase its profits, but concealing the source of support could well give rise to doubt about the results. Advise your students, therefore, to have all co-authors make the appropriate statement clearly in the submission letter or its attachments, leaving to the editor's discretion whether the information is passed on to reviewers or not.

Permission to cite personal communications

Also to be included are any letters of agreement from authors whose unpublished observations have been included in the text as "personal communications". These should have been obtained by correspondence long since, during preparation of the first draft (Step 13). Other permissions may be required: from the subjects of case reports, anyone identifiable in illustrations, and those acknowledged in the Acknowledgments section. All these documents must be included.

Submitting the article

The final submission involves minor but important clerical details: checking that all the numbered pages, plus all tables and figures, are present in each of the copies required, and that an identical set is retained for the submitting author's files. There must be a covering letter to the editor signed by one of the authors only. It is best to send the packet by some form of registered mail, to ensure that the work so painstakingly put together will arrive safely at the correct destination.

It would be strange if at this stage you are asked the naive question "May we submit the article to more than one journal?" If you are, assume that the student missed your exhortations at Step 4, and use the information and arguments on p 15.

"When should I start to worry, if I haven't heard anything?"
A card of acknowledgment should be dispatched by the editorial office on receipt of the manuscript. If this has not arrived within the next 3 weeks, a telephoned inquiry is in order. If it is not the policy of that editorial office to send such a card, authors should politely inform the office that this is in fact accepted practice and a matter of common courtesy.

A decision on acceptability takes much longer. Some journals undertake to make a decision in a certain number of weeks or months, and will say so in their Information for Authors. Most, however, can take up to 6 months. An inquiry after 3 months of silence is not unreasonable, and may jolt the editorial office into realizing that they have not sent the customary reminder letter(s) to the reviewer(s) of the paper. But the best advice you can give to your students is to try to forget about that submission until the 3-month period is up, while getting on with the next piece of scientific work.

Summary of Chapter 8

Informal review, final draft, and submission

- Describe three types of readers who can provide critical review at Step 18:
  - co-authors, and scientists in the same subspecialty or using the same novel methods
  - scientists in the same broad field
  - non-scientific friends and relations

  and require that these all be given a manuscript to read that includes all illustrations and references.

[contd]

- Advise students to specify that the comments should be *in writing*, and to thank the reader before considering and taking any action on the comments.

- Advise students to study the Instructions to Authors thoroughly before revising the paper in accordance with the comments received.

- Give students a checklist (if the target journal does not do so) of actions to be taken in submitting the paper, including the following imperatives:
  - enclose a dated and signed agreement of all co-authors to publication of this paper
  - enclose a statement of any commercial interest on the part of any co-author(s)
  - enclose written agreement to all personal communications by the person(s) quoted
  - check completeness of each copy of the manuscript
  - retain an exact copy of what was submitted.

- Describe the usual time scale of responses from the editorial office.

9
Responding to the journal editor

OBJECTIVES:
To reveal the multiplicity of procedures undertaken by journal editors and reviewers.

To give and explain rules for responding to the editor's decision letter, depending on its tone and content.

YOU WILL NEED:
Copies of Table 9.1 (p 68), with or without footnotes, suitably adapted if you have further or different relevant information.

If available, copies (or an overhead) of an editor's letter requesting revision of a submitted paper that needs to be interpreted as a near acceptance or a near rejection.

YOU MIGHT ARRANGE (though this is optional):
A guest lecture from a journal editor or regular reviewer for an appropriate journal on the kinds of criteria used by reviewers, as specified by the journal or recognized as common practice.

See also "Guidelines for reviewers" in Lois de Bakey. *The scientific journal: editorial policies and practices.* St Louis (MO): C V Mosby; 1976, or indeed the guidelines issued by any broad-spectrum journal in the students' field. If you don't have these yourself, or cannot get them via a faculty colleague, most editors-in-chief of such journals will be prepared to send you a copy if you explain what you want it for.

The arrival of the editor's decision letter is inevitably a moment of great excitement — hardly ever, however, one of unalloyed joy, for outright acceptance without change is rare. It may help your students to become acquainted with the process the manuscript has gone through, and to appreciate the many steps that have led to the editor's decision as well as the several other duties of the editor and editorial office, if you show them Table 9.1.

Table 9.1 Duties of a journal editor and the editorial office

Handling of papers submitted for publication

Acknowledge receipt
Possibly, reject immediately as outside scope
Otherwise, select two or more reviewers, using agreed criteria
Send formal instructions to reviewers[1]
Receive reviewers' recommendations[2]
Make the decision (are further reviewers needed?) and compose the decision letter, enclosing reviewers' comments, edited if this seems necessary
If the decision is acceptance with only minor modifications requested, add editorial suggestions to the copy that is returned to the author[3]
Consider the resubmitted, revised manuscript and decide whether to send it again to the same or other reviewers
Inform the reviewers of the final decision, and copy each reviewer's comments to the others
Send accepted articles to the copy editor, who edits for consistency and "house style", corresponding with the author to obtain final approval, and forwards the result to the printer
When first proofs with the author's markings arrive, reconcile these with those of copy editor and printer's proofreader
Check final proof against the marked first proof

Other duties of the editor

Periodically, engage in major negotiations with the printer
Fix series of deadlines for each issue in the coming year
Make arrangements for author and subject (or key word) indexing of each volume
Make judicious appointments (after consultation) to ensure a balanced editorial board
Maintain a list of reliable reviewers in subject categories
Make and change editorial policy (for instance, on the degree of detailed editorial help the editor will provide; the acceptance and positioning of paid advertising in the journal), in discussion with the editorial board
Maintain good relations with the editorial board, the controlling learned society if relevant, the publisher, the printer
In some organizations, be aware of the financial status of the journal and of tax and charity law[4]

Notes to Table 9.1

1. These instructions, or guidance notes, invariably ask reviewers, as a minimum, to distinguish between (a) detailed criticism to be sent to the authors and (b) the major recommendation on rejection or acceptance with or without conditions, in a covering letter for the editor's eyes only. However, they may be far more detailed than this.
2. Point out here that reviewers are not paid, which explains why it is difficult for editors to impose strict deadlines for review. Discuss whether busy reviewers should be authorized to pass the reviewing task to others in order to speed the process, whether reviewers should always remain anonymous, and other matters the students are interested in. You may wish to call on a colleague who edits or reviews regularly for a journal to attend a session to deal with these.
3. Only if in accordance with journal policy; see under "Make and change editorial policy" later in table.
4. For instance, if costs are partly offset by the payment of page charges, each article must be labeled (discreetly, in a footnote) an "advertisement".

When your students appreciate the number and weight of the decisions the editor-in-chief of a journal has to take, they will be in a better position, and a more sympathetic frame of mind, to take Step 22.

Step 22 Analyze the editor's decision letter and respond appropriately

The editor's letter will contain one of three types of decision:

- outright acceptance,
- firm rejection, or
- request for revision.

There is, of course, nothing further to be done in response to that rarity, *acceptance without modification*, except to break open the champagne, write a grateful note of acknowledgment to the editor, and await the arrival of proofs.

There are four possible responses to a *definite rejection*, and which one is chosen depends on the reasons given.

1 *Send the paper essentially unchanged to another journal.* This should be chosen only if the first editor indicates that the paper belongs in a more, or less, specialized journal. This can only mean that the choice of target journal, taken so carefully in Steps 5–7, was in error, or that the journal has acquired a new editor-in-chief or editorial policy as to its purpose and scope. Even then, it would be unwise to send the paper completely

unchanged, for (as I emphasized earlier) good communicators keep their audience constantly in mind — so at the very least, the author should re-read the paper from the point of view of the new audience and make adjustments where they seem necessary. Authors should also bear in mind — and this is a bitter pill to swallow — that a decision letter advising that the paper belongs elsewhere may be putting into kinder terms the reviewers' opinion that it is not up to the standards of the target journal. This sobering thought may send the authors back to the bench or desk to obtain more data and strengthen the evidence for their conclusions.

2 *Modify and send to another journal.* This is a more likely course of conduct, and will be aided by the detailed reviewers' comments.

3 *Do more research,* and incorporate the earlier results into a new paper at a later stage when deficiencies pointed out by the reviewers have been rectified. The new paper is probably best sent to a different target journal, and will therefore be rewritten with the new journal's readers in mind.

4 *Challenge the decision.* Time and energy spent in arguing with the editor and the reviewers are usually wasted. Only if the authors believe that the reviewers have completely misunderstood a central point in the results and their interpretation, or that they have been unjustifiably critical or simply ignorant of work that supports the authors' contentions, is it worth pursuing this line. Once three or more independent scientists have painfully come to a negative conclusion about a particular piece of work it will be difficult to dissuade them from it, and the goodwill of (probably) senior and distinguished members of the profession and specialty will hardly be gained in the process.

The third and most frequent type of decision, *request for revision,* must be carefully scrutinized as to whether it is (a) an acceptance provided revisions are made in accordance with the reviewers' comments, or (b) an offer of further consideration of a revised manuscript, with no guarantee of favorable outcome. On this will depend whether the authors return to the target journal with a revised manuscript, or send the latter to a different journal where they may have a better chance of acceptance. Before taking either course, the authors should seriously consider revising the paper in conformity with the reviewers' comments; this will also help determine their course. If the decision is to return to the original journal, the authors must supply in a letter to the editor a detailed list of their responses to each of the reviewers' comments. If they decide to approach another journal, the paper must be reconsidered as a whole from the point of view of the different readership.

Above all, urge your students to receive the reviewers' comments gratefully and in a constructive spirit. Heaven knows, it is better to have one's mistakes or gaps pointed out before publication than afterward, and far better to look back on a modest scientific output with satisfaction than on a large bibliography with embarrassment or dismay. Some attention may usefully be paid to the form and style of authors' replies to the reviewers' detailed criticisms, especially if your students' first language is not English.[1] The response should be polite, tactful, grateful for the detection of errors, conceding a lack of clarity if the reviewer has misunderstood an argument or statement, firm but still respectful in rebuttal where the reviewer is mistaken, and with an attempt at writing with the same or a greater degree of formality than the one adopted by the reviewer.

Reference

1 McNab SM. Readable responses to referees. *Eur Sci Editing* 1997;23(3):78-80.

Summary of Chapter 9

Responding to the editor (Step 22)

- Reveal to your class the extent and weight of an editor's responsibilities.

- Describe at least some aspects of reviewers' roles and motivation, possibly with the aid of a guest lecturer.

- Discuss the appropriateness of the following responses to different editorial decisions regarding a submitted paper:
  - *Acceptance without modification*
    Thank, and await proofs.

[contd]

- *Firm rejection*
  Send the paper unchanged to another journal.
  Modify and send to another journal.
  Do more research.
  Contest the decision.

- *Request for revision.* Decide whether it is:
  Acceptance contingent on suggested changes,
  or only an offer of reconsideration after major change.
  Act accordingly.

- Advise authors to accept reviewers' comments in good part, in the spirit in which they are (usually) offered.

10
Conference posters and their abstracts

OBJECTIVES:
Using the same series of steps as in writing a journal article, to reveal how similar the two processes are.

To emphasize the need, in a poster, for visual impact and the suppression of unnecessary detail.

To describe what to take to the conference and how to discuss the poster with viewers.

YOU WILL NEED:
Copies of Table 1.1 (p 9).

ASSIGNMENTS:
Prepare a poster abstract and the corresponding poster as the major assignment, as an alternative to preparing a journal article.

Minor assignments as in Chapters 1–7.

TIMING:
Simplest, perhaps, as a separate 1-hour session at the end of the course on writing a journal article. But if a number of your students are preparing a poster as their major assignment, you may have to intersperse some instruction as given below at the relevant steps.

Increasingly, major conferences are gaining excitement and topicality by inviting participants to offer poster presentations, sometimes as a condition of attendance. It is not unusual for institutions to fund researchers' attendance at a conference only if their poster presentation has been accepted.

For young and as-yet-unknown researchers, attending conferences is a valuable way of learning what is happening in the mainstream — and the tributaries — of their field. The preparation of a poster, for discussion in person with other workers at the conference, is a valuable exercise in scientific communication.

Posters usually present work in progress or a set of results that form part of a larger investigation, as yet incomplete. Much of what is recommended in the preceding chapters about preparing a journal article applies to the preparation of a poster, apart from the requirement of completeness, and of course with much of the writing condensed and special emphasis on "the visuals". You may have allowed preparation of such a poster as the main assignment in the course, provided the student devotes ample thought to the necessary differences between the two forms of communication.

Preparing a conference poster usually comes in two parts, namely (a) preparing the abstract, sent to the conference organizers ahead of time and subjected to peer review, which may determine whether the poster is accepted for presentation at the conference or not; and (b) preparing the poster itself.

As you will be teaching those who either have previously attended the course on preparing a journal article or are attending it now alongside students whose major assignment is the journal article, the easiest approach is to use the same table of procedural steps (p 9) and adapt it for this somewhat simpler task. The scale of the undertaking is so much less than that of a journal article that some steps can be omitted or quickly passed over (e.g. Steps 8, 10, 14, 19, 20, 22). The physical preparation of the poster (corresponding to having the illustrative material professionally made, Step 15 in preparing a journal article) comes quite late here, namely after submission of the abstract at Step 21, but its over-all design will be at the forefront of the planning from Step 11 onwards.

Step 1 Ask yourself whether the time is right

This step is less relevant than for a journal article, as posters often present *incomplete* results precisely in the hope of stimulating helpful one-to-one discussion with participants in the conference who have valuable relevant experience. But newcomers to a field should probably steer clear of presenting results that are so preliminary as to be open to a wide range of interpretation, or subject to possible retraction after further work.

Step 2 Clarify your conclusions by sketching tables and figures complete with titles and footnotes

Step 2 is even more vital here than for a journal article, since (a) thinking hard about the illustrations will (as with a journal article) clarify the logic of the approach and the validity of the conclusions reached to date, and (b) it is the illustrations that will convey the main message in the eventual poster.

Note, however, that many conferences specify that there must be no tables or figures *in the abstract*, so that (a) helps in writing the abstract, and recognizing (b) helps in preparing the poster itself.

Step 3 Decide who will be co-authors

This may be important if others wish to share the priority of the announcement, or if the young postgraduate in floating his or her first publication into the scientific pond wants to show in whose department the work has been done, and hence wants to add the name of the head of department. A better way of doing this is to add the head's name after the name of the department in the affiliation line, or in an Acknowledgments section. Even more important is to decide who is actually going to stand in front of the poster and defend or discuss it with viewers. This is the one who should be given main charge of composing the poster and writing the abstract.

Step 4 Consider the ethics of scientific publication

Step 4 arises because some of your students may have been told that publishing a poster abstract precludes subsequent publication of an article containing the same results. Disabuse your students' minds of this fear: prior publication of a mere abstract is not a deterrent to publication. It is good practice to inform the journal editor that the poster abstract has previously been published or circulated when submitting the full paper (which will usually contain far more results than did the poster), and it is usually appropriate to cite the abstract — labeled as such — as a footnote on the opening page and in the references.

Step 5 Relate your conclusions to the existing body of knowledge

The interests of the majority of participants in the conference must be borne in mind when the general approach is planned. If the student has picked a conference because it is in an attractive locale and not for sound scientific reasons the poster runs a considerable risk of rejection on grounds of irrelevance.

Step 6 Write a working title and abstract

Step 6 always constitutes excellent mental exercise, as it focuses the mind on the precise subject and range of what is to be presented. Usually, conference organizers impose a limit on the number of words in the final abstract; the limit need not be strictly observed at this stage, whose purpose is personal, but should not be greatly exceeded either, as a poster should not attempt to convey more than one or two messages.

Step 7 Choose the target journal and make notes on its Instructions to Authors

The "target journal" is irrelevant, but the advice to make notes on the Instructions to Authors, this time issued by the conference organizers, must be diligently followed, as the constraints on poster preparation are often tightly restrictive, and should govern the whole process.

Step 8 Name the main sections

This step should be skipped, as here we part company from the journal article. The IMRAD structure is rarely appropriate for poster presentations: with posters, it is the main conclusion that needs to catch the viewer's attention and therefore must be prominently announced, perhaps first in the title and again elsewhere in visually arresting form, quite out of the usual order. The following structure is one for which the viewer, who may be walking past 100 or more such posters, will be grateful:

- Rationale for doing the work
- Results obtained (shown strikingly, if possible in such a way that the conclusion can be instantly drawn)
- Significance of the findings.

Step 9 Fill section files with relevant brief notes in any order

The number of items added to the ragbags should in this case be pared down to the barest minimum. Unless the poster is about a novel method, methods should be summarized, not detailed. The poster presenter should of course go to the conference armed with full methodological details as well as any supplementary results that may be needed to answer viewers' questions, but the poster itself should be as simple and clear as the author can achieve while still making scientific sense.

Step 10 Arrange the contents of section files logically

Hardly necessary, if the contents have been reduced to a small number of items.

Step 11 Finalize the design and content of tables and figures

From this point on, the poster presenter should be thinking in terms of the layout of the entire poster, with eye-catching illustrative material as the key to the whole design. The poster should not be made yet, as it will not yet be known whether its presentation has been agreed.

Step 12 Make a topic outline

In making a topic outline, the author should submit to the discipline of confining the poster to three clear statements leading to a single conclusion — or a question. A sentence outline is not needed, as it would constitute the entire text.

Step 13 Write the first draft continuously, collecting references as you go

This should be easy, if the author starts from the illustrative material, mentally placing it on the poster and sizing it in such a way that it is easily read and comprehended, then fitting a minimalist text around it in accordance with the topic outline. Vital references should be noted as writing proceeds, and re-read in their entirety before anything further is done. They should not, however, clutter up the eventual poster but should form part of the armamentarium of the poster presenter on the day.

Step 14 See if the first draft needs major alterations

Should be unnecessary in such a short work, unless re-reading the crucial references has modified the author's opinion.

Step 15 Have illustrations professionally made

This again differs from preparation of a journal article. The entire poster may eventually need to be professionally designed and printed, or at least prepared with professional advice and help. The author should defer preparation of the final poster until the abstract has been submitted (Step 21).

Step 16 Polish the prose

The words of a first draft can always be edited to advantage. Apply the editing techniques outlined in Chapter 7.

Step 17 Rewrite the title, and structure the abstract

Both of these must be made as appealing to the reviewers as possible. Whether or not there are restrictions on the length of the title, it should be as short as is compatible with conveying the concepts accurately. In all poster work, authors must constantly remember that they are competing for viewers' time in a sea of other posters, and they must use every technique they can muster to be arresting.

The abstract too must vie with others to command attention and lure the reader to view the poster itself. Sometimes, the Instructions to Authors demand a specific structure for the abstract. If they do not, advise your students to make the abstract as short as possible, shorn of unnecessary detail, and to use strong verbs in the active voice.

Step 18 Request informal, friendly review by three critics

Review of the abstract by co-authors, local colleagues, and mentors experienced in going to conferences should yield useful tips for revision of the abstract and in due course of the layout of the poster as well. The non-scientific friend is less useful in reviewing posters, as poster content is so succinct and is specifically directed to a specialized audience.

Step 19 Re-read the Instructions to Authors and make any necessary adjustments

No adjustments should be necessary.

Step 20 Revise up until the final draft

The advice of co-authors and (or) experienced local colleagues should be sought on the layout of the final poster, but the presenting author should have to make no more than one revision of the original sketch.

Step 21 Submit abstract to the conference organizers

Remind students that if it is a large conference with many subsections they must specify under what subject heading they wish the abstract to be considered and in what subject section the poster, if accepted, is to be placed. Rigorous attention to the instructions to participants is needed here.

Construction of the poster whether used or not

Once the abstract has been sent off, construction of the poster itself may begin. Even if, unluckily, the presentation is not accepted the poster can usually be used to good effect for local seminars or departmental Open Days, so the time will not have been wasted, and the student will have gained valuable experience to be put to good use at another time.

Presentation of the poster

Here is a checklist of what the poster presenter must take to the conference:

- the poster
- detailed account of the experimental procedures omitted from the poster
- further copies of the tables and figures
- any later results obtained since the poster was made
- list of references with full bibliographic details
- calling cards bearing name, address, and contact numbers of the home institution.

"Presentation" usually consists simply of standing in front of the poster (to the side, not blocking the view!) and answering questions, or prompting them from those who look interested. At some conferences, posters are formally presented in 10-minute slots, in which case the advice in Chapter 11 should be heeded.

Summary of Chapter 10

Preparing conference posters and their abstracts

- Propose the same series of steps, only slightly modified, as for preparing a journal article, advising that Steps 8, 10, 14, 19, 20, and 22 can probably be omitted, Step 15 should be postponed until after submission of the abstract, and some steps can be shortened.
- The poster abstract, to be sent to the conference organizers for peer review, and the poster itself have to be conceived and designed together.
- Emphasize that a conference poster should aim to make only one or two points.
- Posters and the abstracts that precede them have to vie with others for the attention of readers and viewers and must therefore be strikingly designed to contain a few strong messages.
- In adapting the series of steps (p 9) in preparing a journal article, repeatedly point out that the end-product must be striking to the eye and quickly comprehended, so that simplicity of expression is of the essence. Achieving such simplicity takes hard work.
- The whole process should take less time than preparing a journal article, but demands just as much mental discipline, especially in eliminating irrelevant items.
- List what poster presenters need to take with them, besides the poster, to the conference.

11
Scientific talks

OBJECTIVES:
To enable students to plan and prepare an effective scientific talk to be given in a specified number of minutes.

To give students guidance on oral presentation at venues of different sizes and types.

To provide opportunities for rehearsal with constructive comment.

YOU SHOULD READ:
Peter Kenny. *A handbook of public speaking for scientists and engineers.* Bristol and Philadelphia (PA): Institute of Physics Publishing; 1982.

YOU SHOULD CONSULT:
Council of Biology Editors Scientific Illustration Committee. *Illustrating science: standards for publication.* Bethesda (MD): Council of Biology Editors, Inc; 1988 (for tips on the preparation of slides).

ASSIGNMENTS:
Transform a complex table into legible slides, with titles, suitable for projection.

Deliver a 10-minute talk to the class under various (possibly simulated) conditions.

TIMING:
At least three 1-hour sessions.

This subject is best taught in three or more sessions. In the first, you expound the principles and give out assignments; in the others, students make 10-minute presentations for comment and criticism by you and by their fellow students. Students may present some aspect of their own work or base this talk on a paper drawn from the literature — perhaps one that they have analyzed at Step 5 in Chapter 2. If you have the facilities for video recording and playback, this can be used to advantage, but there is no substitute for *immediate* feedback from a live, critical but sympathetic audience. If you can arrange to have that audience seated toward the back of a lecture theatre with poor acoustics and weak amplification, so much the better. Perhaps these extra impediments should be reserved for the second session of presentations by students, after they have gained some confidence under less trying circumstances.

Your own guidance comes under three headings: the planning, writing, and delivery of a talk.

Planning a talk

First, lay out the over-all aims that students must bear in mind from the start:

- Capture audience attention at once.
- Clarify your main objective.
- Cut down strenuously on detail.
- Use short sentences without subordinate clauses.
- Use simple slides.
- Don't separate results and conclusions.

The main object of planning is to provide a strong framework, a clear structure for the talk. Advise students to use the simplest structure for their first attempts: *Why* you did the work, *How* you went about it (without detail), *What* resulted, and *So what?*

In a talk longer than 20 minutes, the audience should be told what the structure is going to be. In a short talk, there is no time for this; it should be obvious from the talk itself.

Accurate timing is vital to successful oral presentation, and you must give firm guidance on timing, and the planning of timing, throughout. Tell your students to aim to talk for a shorter time than that allotted, and thus avoid the need to double their rate of delivery in the last 5 minutes, which has the unfortunate result that the most important part of the talk, the summary and final message, is lost in the rush. Even with all the careful planning outlined below, one must still allow a little time to cope with the disasters that occur with surprising frequency: malfunctioning of the lighting, microphone, slide projector, or laser-light pointer, upside-down slides, and disturbance within the audience or outside.

Writing the talk

Next, insist that students prepare their talk in writing. Not only does this reveal that everything they begin by wanting to say cannot be intelligibly delivered in the time allotted (especially if this is only 10 minutes, typical of what junior scientists are given), but the written word also enables one to

- check logical development
- ensure good transitions from one thought to another
- check that sentences are short
- select vigorous verbs
- drop inessential items
- devise lively phrases
- check that only familiar terms are used
- allocate balanced amounts of time to the different sections.

As with a journal article, it is best to begin with the visual material: the slides, or transparencies for overhead projection. Having selected what material to present, students must work out exactly how the slide or transparency should look. Tell them to

- begin by allocating to each potential slide a title — this will enable listeners who have dozed through the preceding sentence to catch up with what the slide is intended to convey;
- be sure that each slide makes only one point, and if possible convey that point graphically rather than in tables of numbers — an occasional text slide can be interposed;
- list the slide titles and examine whether they make a logical sequence;
- reckon on showing slides for not more than 80% of the time allotted, at 1 minute per slide — thus, more than eight slides in a 10-minute talk is too many.

Having selected the slides for minimal number with maximal effect, the student must plan exactly what to say about each. It is pointless, and irritating to the audience, to read off the words on a slide as though the listeners are blind. The words should complement, illuminate, and enliven the slide.

The introduction

Once one has a structure, a skeleton for the talk (with provisional timings), one can begin to flesh it out. Get the students at this point to tackle the fun part: preparing the *approach to the audience.* The first words of a talk can make all the difference between success and failure; the object is not only to capture the audience's attention but to make the audience sympathetic and receptive to the message. Here are some possible openings:

> *Narrative*: "Last year, someone in my lab/institution/country made a discovery . . ."
> *Quotation*: "James Watson once said . . ."
> *Rhetorical question*: "Would anyone deny that cholesterol is a killer?"

Startling statement: "Global warming is, in my opinion, nonsense."
Negative statement: "Global warming is not going to be a problem."
Comparison or contrast: "Arizona has one-tenth the average annual rainfall of Oregon. Yet . . ."
Statement of common interest: "We are all concerned about culling polar bears. Our recent research has focused on . . ."

There is no scope for a lengthy introduction in a 10-minute talk, so the approach has to be made in a single sentence. But that sentence deserves a lot of attention: it can set the tone of the whole talk. A longer talk can stand as much as a paragraph. Tell your students not to waste time in either case thanking for being invited to speak, unless they have something extremely witty to insert into the thanks: a conventional "thank you" on this score always sounds insincere, and is boring besides. Above all, they must not spend precious minutes apologizing that there is insufficient time to cover the subject (yes! I've actually heard such apologies, and they were lengthy).

If none of the devices listed here can be made neatly relevant to the talk that follows, they should not be used. An irrelevant introduction that leads the listeners' minds in the wrong direction only confuses. And an introduction consisting of a would-be funny story that falls flat is the worst of all.

The body of the talk

In writing the main part of the talk — wording the sentences to be spoken about each slide, and any sentences needed to link them — students should follow the same style guidelines recommended in Chapter 7. Thus, they should use short, clear statements (Style Guideline 1) and words correctly related to one another (SG2) while avoiding abstract nouns (SG3), the passive voice (SG1), and strings of modifiers (SG4).

Much attention must be bestowed on the technicalities of actually making the slides that have been planned. Poor slides are notoriously the cause of exasperation for audiences. Table 11.1 shows some do's and don'ts.

Table 11.1 Making slides

DO	DON'T
Use standard size (35 × 35 mm)	Use illegibly small type
If possible, fit material in a rectangle, the long side across	Use poor-quality prints plus an apology
Use colors for emphasis	Make slides from poorly printed copy — have them re-lettered
Put arrows on micrographs to show point(s) of interest	

If the graph on a student's slide absolutely must have several lines, these should be differentiated by color coding, only colors that contrast well with the background being used. Axes on graphs must be labeled, otherwise they are meaningless.

> ASSIGNMENT
> Transform each table and graph you used for a journal article, or a complex table from a published article, into one or more slides, complete with titles, that will make the points easily comprehensible to a listener at the back of a large auditorium.

The summary

The concluding sentences are as important as the introduction for ensuring success. The conclusions and significance must be summarized clearly and succinctly, and the final words should encapsulate the "take-home message". These words are the ones most likely to be remembered afterward, and so should be among the brightest and best. One-tenth of the total timing should be allowed for delivery of the summary.

Deciding whether to read the talk

Ideally, talks should not be read. Communication by talking comes off best if there is eye contact between the speaker and those spoken to, and eye contact is impossible to achieve if the speaker's eye is glued to the text, with occasional nervous glances at the screen.

Canny speakers who find it difficult to memorize whole talks design their slides in such a way and in such an order that their appearance automatically prompts the words that must be spoken next. Others make do with short notes on index cards that can be slipped into a small bag or pocket. The contents of these notes depends on the speaker's preferences. For some, a topic outline is enough. But most speakers need a reminder about the transition from one topic to the next, and some may need to be reminded of the key phrase or sentence that starts or ends a topic. There is no single piece of advice helpful to everyone; tell your students to experiment, and when it comes to the sessions in which the students "perform", ask which method they chose and why.

Having said that, ideally, a speaker should not read the talk, I hasten to admit that some people can manage quite successfully to read their text while maintaining eye contact. Their text must be printed out double-spaced (or 1½-spaced), sometimes with enlarged print, in strong black type. The page turns should be managed so as to correspond to natural breaks or pauses. The speaker will need to know the text very well (this comes from several rehearsals; see below) and therefore to be aware of the intended effect of the sentence just embarked on and what is coming next. One tip may be borrowed from frequent public speakers: instead of a period at the end of each sentence, they print a series of dots to remind them to pause, briefly, and not to rush on into the next idea.

A problem with reading aloud is that the speaker is not thinking out the sentence while it is being spoken, and may therefore place the stress in the sentence inappropriately, which impedes the listener's comprehension. The speaker must also be skilled enough to glance up frequently, to different parts of the audience or to the screen, without losing the thread of the sentence or the argument. But all these things can be acquired through practice, and if reading the talk, after plenty of practice to avert the known hazards, makes all the difference between a performance wrecked by fear of memory blanks and a performance marked by quiet confidence, so be it. Whether the performance is a success will probably depend less on the mode adopted than on the conscientiousness with which rehearsals are undertaken.

Rehearsal

After preparing the text and slides, the prospective speaker must rehearse the delivery, if possible several times, in front of different listeners who are prepared to give constructive comment. The first rehearsal can be for oneself and enables the speaker to judge

- whether the timing is about right,

- whether there are any tongue twisters inadvertently written into the text,
- whether the sentences are indeed short and clear enough, and
- whether there are long periods without slides, followed by a burst of several slides in quick succession; whether this is desirable and, if so, whether to ask for lights during the blank periods.

The second rehearsal might be in a large room at home, perhaps with an audience of the speaker's or a borrowed cat, who will probably sleep through part of the talk and thus nicely simulate an audience whose attention has not been captured.

The third rehearsal ideally takes place before one or more colleagues in a lecture room, the future conditions being mimicked as closely as possible. Before this final rehearsal, or mock performance, the points made under "Delivery" below should be mastered. You can take your class through several performances of this kind, under different circumstances as described in the opening paragraph of this chapter.

Presentation of a talk at these sessions constitutes the major assignment for this short course. You will have to overcome your students' natural shyness and reluctance to perform before their colleagues, but persuade them that this audience will be kinder and more inclined to give constructive criticism than those at distant conference venues; and you will find that after their ordeal the students will thank you for having forced them into it, as they will have learned a lot that "their best friends wouldn't otherwise have told them". Ask your shyest and most introspective student to give this practice talk, just as an experiment, in an over-emphatic, theatrical way, with a degree of projection the student considers quite inappropriate, and ask for the class's reactions. If the room in which it happens is any larger than a classroom, the chances are that the class will find it a refreshingly clear performance, not in the least exaggerated.

It is worth insisting on the point that rehearsals, both strictly private and then before a small, sympathetic audience, are crucial to successful public speaking, which comes naturally to few. The time for rehearsal must therefore be taken into account in calculating the total time needed to plan and prepare a talk.

Delivery

Your students will find the following tips useful, whether they are totally inexperienced speakers or long-suffering listeners. It helps if you impart the tips with a degree of humor.

- Go suitably dressed to the session in which you are to speak. This means that you should not be dressed more casually than the prevailing dress code, if anything a shade more formally. Extremely informal clothing will not be commented on, but is disrespectful to your audience and could subconsciously antagonize them.

- Before you set out, check that you have your slides, your notes, and the full text with you and keep them on your person at all times.
- Ensure that all the slides are present, right way up, and numbered ready to be loaded into the cassette.
- If possible before the session starts, and if not during one of the earlier talks, locate the pointer, microphone, and remote-control buttons for advancing the slides.
- If a clip microphone is used, find somewhere on your clothing where it can be held only a few inches below your chin: this may affect your choice of clothing in the first place. Make sure the wires (if there are any) are so positioned as not to trip you up. If there is a fixed microphone, adjust it to the level of your mouth and project your voice into it from a distance of about 15 cm. You must not permit yourself to move far from that spot even when turning to point at the screen.
- Adopt a relaxed but not sloppy stance — no hands in pockets or strings of pearls! Stand with feet about 30 cm apart, one foot slightly advanced, to the side of the screen. Seize the pointer in the hand nearest the screen and either drop the other to your side or rest it lightly (not with a white-knuckle grip) on the lectern if there is one.
- Start with a warm smile at the audience, and launch straight into your first, carefully prepared and memorized sentence. Do not begin with cumbrous acknowledgments to all your co-workers. They will have been listed in the prior publicity.
- Don't gabble. Speak deliberately rather than slowly, allowing just enough time between sentences for the audience to catch up with your thought. At the end of paragraphs, use the three-dot technique (p 86), whether you are reading the talk or not.
- Use gestures sparingly but naturally. The most natural gesture is to point at the screen; do this with a fluid movement, returning as soon as possible to face the audience. Do not talk to the screen for long.
- This is where transparencies on an overhead projector (which are preferable to slides for small audiences) have the advantage: you can point with a pen to any part of the image without taking your eye off the audience, since you do not have to turn round to see what is appearing on the screen. Do not be tempted, however, into leaving it to the last minute to make the transparencies; creating them always takes more time than

you expect, and it is rude to your hosts to take some of your talk time in doing it. An exception is if you plan some interaction on the transparency with your audience.

- If there is a mishap, pass it off with an unrehearsed quip. However feeble your remark, if it is swift the audience will laugh sympathetically, and this will further relax you and them.
- Never apologize. If your slides are unintelligible you should not be showing them. If you have not had time to complete your talk you should have stayed up all night doing so. If your experiments have not lived up to your expectations you must present what facts you have and draw what conclusions you may. If absolutely necessary, withdraw your contribution, giving the chairperson as much notice as possible. Presenting a pointless talk with an apology and sickly smile is an insult to any audience, for you have effectively wasted their precious time.
- Dim the lights, or ask for them to be dimmed, only as you approach the first slide. The audience likes to see you for at least the opening sentence.
- If the pointer projects a small red arrow at the screen, either switch it off or point it to the floor between uses; do not allow it to flicker around the room while the audience follows its course like kittens watching butterflies.
- Show each slide long enough for the audience to take in its significance, but don't leave it on ages after your talk has moved on. If there is a planned long gap between groups of slides, turn on the lights, but do not keep flicking lights on and off.
- Remember not to read aloud every word on the slide.
- Deliver your rehearsed and memorized peroration in a particularly firm voice, not dropping the tone at the end of the last sentence but enunciating the final consonants clearly. Then turn toward the chairperson to signal you have finished. It should not be necessary to say "Thank you (for your attention)" if you have timed things well, but there is no objection to doing so. Do not, however, ask for questions. This cuts off any applause that might be coming your way; anyway, questions are at the chairperson's discretion, not yours.

After you have delivered these tips and the students have duly noted them, name the students who during the next session will make presentations as though at a scientific conference. It is of little use to wait for volunteers. Make the ambience for the first session comfortable and not threatening.

At the first performance session it is best for you to lead with your criticisms, or rather suggestions for improvement, with reference to the previously delivered tips; only after you have set the tone of constructive suggestion invite the other students to make their comments — they will almost certainly have noticed features you hadn't, and may have other good tips to add to yours. In subsequent sessions you can generate more difficult conditions and encourage more audience participation. If any performances are really awful, cut discussion fairly short and offer some private tuition, or "extra practice", in which specific individual problems can be tackled. This will undoubtedly be welcomed.

Summary of Chapter 11

Planning and delivering scientific talks

- Teach the three necessary aspects of giving a talk:
  planning, preparation, delivery.
- Emphasize the importance of timing, and the planning of timing.
- Discuss ways of introducing and summarizing the talk.
- Give do's and don'ts of slide-making.
- Insist on the need for rehearsals.
- List a number of tips on good presentation, then take your students through several performances under increasingly trying conditions.
- First make your own suggestions for improvement of each presentation (gently!) before the whole class, then gradually allow or encourage other students to participate to a greater extent.

12
Dissertations

OBJECTIVES:
To draw parallels between writing one or more journal articles and writing a dissertation, in particular emphasizing the cardinal principles of logic, clarity, precision, and relevance.

With due deference to the awarding institution's instructions to candidates, to dissuade students from spending most of the available time on constructing an exhaustive historical background, pointing out the virtues of judicious and discriminating selection from previous work.

BACKGROUND READING OR REFERENCE:
E S Balian. *How to analyze, design and write doctoral research — the practical guidebook.* Washington (DC): American University Press; 1982.

W C Booth, G G Colomb, J M Williams. *The craft of research.* Chicago (IL): University of Chicago Press; 1995.

Books on searching the literature in the Annotated Bibliography (p 190).

The following two books also contain chapters on writing theses:
Maeve O'Connor. *Writing successfully in science.* London and New York: Chapman & Hall; 1991.
David Lindsay. *A guide to scientific writing.* 2nd ed. Melbourne: Longman; 1990.

YOU SHOULD CONSULT:
A research supervisor in your institution if you are not yourself acting in this capacity.

Mimi Zeiger. *Essentials of writing biomedical research papers.* New York: McGraw-Hill; 1991, for further exercises on style at Step 16.

YOU WILL NEED:
Copies of Table 1.1, p 9.

TIMING:
A single session, toward the beginning of the students' research period, should be devoted to methods of searching the literature and compiling an annotated bibliography to guide the planning of the research. A select bibliography for your guidance on these matters, if you decide to teach such a session, is included on p 190.

This chapter, however, is concerned with teaching the actual writing of the dissertation about a year before the end of the research period, and may conveniently be given in a few sessions at the end of the course on writing a journal article described in Chapters 1–9. At that point the students will have become accustomed to a logical step-by-step procedure, which is also recommended here. The session(s) can take the form of a question-and-answer seminar in which the students themselves devise adaptations of the steps in writing a journal article, p 9, and end up with something resembling Table 12.1, p 101.

This order of events works well for most students, who will almost certainly in the course of their research have at least *participated* in preparing one or more journal articles and have learned how to draft one. They will therefore be in a good position to appreciate the similarities and differences between the tasks, and will be able to tackle each of them with equal aplomb.

A dissertation is generally longer than a journal article, but it need not take proportionately longer to write. Perhaps the best advice you can give to students is: *Aim to make your dissertation the shortest that the examiners have ever awarded a distinction to.* Many students set out believing that their dissertation must contain more than some arbitrary number of thousands of words, and this thought becomes an intolerable incubus lying heavily on their spirits throughout their days as a graduate student. Disabuse their minds of this notion, and encourage them to move calmly through a series of steps similar to those recommended for writing a journal article, aiming in the end to be succinct and incisive.

Invite your students to examine the steps in Table 1.1 in class, specifying which of them applies, with or without adaptation, to the dissertation. Some complementary points that may not emerge in class are outlined below. But first, a quick run-through of the differences between dissertations and journal articles. This chapter makes no distinction between theses and dissertations, terms that are used differently and interchangeably in different institutions. The advice relates chiefly to PhD dissertations; modifications for other degree or diploma requirements can readily be imagined or ascertained.

A few differences between dissertations and journal articles

The most obvious difference is that a dissertation is the work of a single author, who has to convince examiners that he or she is qualified to pursue an independent career in scientific research, has learned how to conduct investigations in a scientific and ethical manner, can critically evaluate the published work of others as well as his or her own work, and has some presentational skill.

A dissertation is longer than a journal article for three main reasons:

- the need to demonstrate mastery of the literature on the subject
- the need, often, to cover more than one topic or line of investigation, sometimes only loosely connected
- the desirability of indicating future development of the author's research, to show the examiners that the student has gained scientific perspective and an appreciation of how new hypotheses are formulated.

Nevertheless, continue to discourage prolixity: competence is best demonstrated through concise and pungent expression.

Bearing these distinctions in mind, your class can proceed to examine the steps in Table 1.1 and to adopt or adapt them in writing a dissertation.

Step 1 Ask yourself whether the time is right

Basic preparation of a dissertation starts early on. An outline of the plan of the research should be written a few weeks or at most months into the research period, with careful selection of headings and subheadings. The outline should be formally reviewed at fixed intervals of a few months throughout the dissertation period, and at each review modified in accordance with the findings to date. This process guides and controls the research, and usefully imposes structure on the eventual dissertation.

Whether the time is right to start writing the complete dissertation for submission may not be a question students can allow themselves the luxury of asking: external constraints on its production include timing, funding, and the requirements of the student's supervisor. The ideal time to write and present a dissertation is when a distinct phase of an investigation has come to an end and when asking a well-defined question has led to reliable and worthwhile conclusions (or to the better definition of further questions). We assume in what follows that this stage has been reached, and writing can begin.

Step 2 Clarify your conclusions by preparing tables and figures complete with titles and footnotes

The rough sketches in notebooks or computer files obtained to date must now be transformed into presentable versions complete with titles, subtitles, column headings, side-headings, footnotes, and legends. Then the titles must be listed in the intended order to make sense to a reader — specifically, the internal and external examiners — without reference to the text in the Results section. Finally, the illustrations *together with their footnotes or legends* must be examined in that order, with the same aim of clarifying, for the author, the logical thread running through the whole of at least one discrete section of the research.

The process can, if critically performed, reveal flawed logical reasoning, forgotten conflicting results, missing or inappropriate controls, insufficient evidence, unlikely statistical conclusions — all matters that can probably be put right if the student embarks on the step early enough. In the best case, it will signal the answer to the question in Step 1 — that the research phase is completed, or will be in a finite number of weeks.

Thus, the main purpose of Step 2 is private: to clarify the author's thinking.

Step 3 Decide who will be co-authors

This step can be omitted: a dissertation is the output of one person. However, now is a good time to remember, and list, those whose help should be mentioned in the Acknowledgments.

Step 4 Consider the ethics of scientific publication

Again, this step does not apply in the same way as for a journal article; the dissertation will not, for example, be hawked round a number of learned bodies to see who will accept it. The ethics of animal or human *experimentation* will, of course, have to be discussed if relevant. It should be unnecessary to say (but you might as well say it!) that the inclusion of other people's data masquerading as the student's own, not to mention *fraudulent* results, is strictly forbidden.

Step 5 Relate your conclusions to the existing body of knowledge

This is where many dissertations come a cropper. Word has somehow got about that every dissertation must begin with a huge, exhaustive survey of preceding literature on the topic. Rumor also has it that the more references over 100 this section contains, the more brownie points the author will earn.

Perhaps the rumor has been fostered by the belief, mentioned earlier, that the dissertation must contain vast numbers of words and that a hefty literature survey is the easiest way to boost the word count.

None of these allegations is true. What singles out a dissertation as particularly commendable is a succinct, cogently argued, perceptive analysis of closely relevant references, judiciously selected from the mass of possibilities. Students worth their salt will of course have read a great deal more than they directly need to underpin their research, and it is natural that they should want their diligence to be noticed; but convince them that a knowledgeable examiner will almost certainly be able to deduce the breadth of their reading from the insight with which they discuss the references they do cite, and will on the contrary mark them down for dragging in the irrelevant and the peripheral. The advice is, then, as for a journal article: keep the introduction short, powerful, and to the point.

Students could fulfill the desire to display their conscientiousness by compiling an inclusive bibliography from which the "references cited" have been drawn. If they are wise, or have been well counseled, they will have been adding to this bibliography, with brief annotations, ever since the research began. Some supervisors advise students to include this bibliography as an appendix to the dissertation. Point out to your students that if they do so they must be thoroughly familiar with the contents of each item, otherwise they could find themselves having a rough ride while a disbelieving examiner tests their acquaintance with intimate details of works they claim to have studied.

Advise the students above all not to get hung up on the literature survey and the introduction at this stage, or they will spend the bulk of the available time on honing it, and then find themselves having to hurry through the most original and interesting part of the dissertation — their own results and conclusions — and failing to do it justice. They should instead concentrate on identifying and discussing the work that led to this particular investigation, and not linger endlessly in the past. Have them move on quickly to the following steps, especially Step 9.

Step 6 Write a working title and abstract

This is again for private purposes, to delimit the topic and mark out the dissertation's boundaries.

Step 7 (Choose the target journal and) make notes on the Instructions to Authors

The first part of this step does not apply, but any Instructions for Authors of Dissertations issued by the university or other awarding body must be obtained and carefully studied now.

Step 8 Name the main sections

Headings for the main sections are often prescribed by the body awarding the degree or diploma, and are usually the same as in a journal article, but some consideration should be given to unconventional possibilities if the material warrants it. If the student has developed subheadings in the course of the research, separate section files should be made for main headings and for subheadings as well.

Step 9 Fill section files with relevant brief notes in any order, to form "ragbags"

Pouring thoughts, concepts, results, key references and stray telling phrases into named section files, or "ragbags", in higgledy-piggledy, non-logical order is a wonderfully liberating technique for unblocking the dams of fear that may have built up during the research while the required submission of a dissertation became an ever-larger cloud in the sky.

Step 10 Arrange the contents of the ragbags logically

The ragbags filled with pieces of information in Step 9 now need to be sorted into logical order, which should present little difficulty. Some exclusion of the more peripheral notions may be in order. This step goes hand in hand with Step 11.

Step 11 Finalize the design and content of tables and figures

Most of this work will have been done at Step 2, but contemplation of what is to be included, in what order, as decided at Step 10, may call for some modification.

Step 12 Make a topic outline and consider writing a sentence outline

Steps 10 and 11 almost automatically lead to a topic outline, which provides the final strong framework for writing the text. As explained in Chapter 5, sentence outlines help some people, not others. But perhaps, as a dissertation is a more extended piece of work than a journal article, you should recommend a sentence outline somewhat more vigorously in this instance, especially to the inexperienced writer: the exercise of repeatedly having to encapsulate a message into a single key or topic sentence for each paragraph is as healthy for the mind as jogging is for the circulation.

Step 13 Write the first draft continuously, collecting references as you go

The advice to aim for continuous production still pertains, even though it is impossible for anyone to write a whole dissertation, however roughly, at a single sitting. Encourage students to start at the beginning of the Introduction and write at least one complete subsection at the first session of several hours; then, at the second sitting, to press forward without attempting to revise the first portion, and so on each time. Students should write steadily forward, keeping the topic outline and any sentence outline before them, making a brief note of relevant references as they go, for prompt bibliographic checking at the earliest opportunity.

It is important at this stage not to get bogged down in any one passage but to keep a continuous, logical flow of ideas going. If a passage proves sticky, the student should revert to note form, without bothering with grammatical problems or stylistic infelicities. If, at the end of the Conclusions, Interpretation, or Recommendations — whatever the dissertation ends with — the writer feels he or she has arrived there by an unnecessarily roundabout or extended route, *then* is the time to think about revamping whole sections, but not before. Otherwise, despair descends as time dwindles away and the second section (say) is in its ninth draft and still incomprehensible while the rest is non-existent.

The considerations within the main sections are pretty much those of the journal article, with the following exceptions.

- The Introduction can be longer and written with the aim of demonstrating mastery of the subject and its background (while remaining succinct!).

- The Methods section can also be more extensive, with critical comment on the methods both rejected and employed.
- The Discussion may include more about the appropriate way forward, analyzed with verve.
- There may be an Appendix describing investigations undertaken but not completed, if it looks as though others may be able to build on this work.

Some words of comfort you can give the students to remember if they become desperate are: "At least you have the whole thing written, if only as a first draft. Now with that in mind, start reading the part you consider the weakest. You will be surprised to see that there are passages you are quite pleased with; analyze the others *either* by reducing the text to an outline, or comparing it with the original outline, and writing the section again in the crispest, brightest way you can, *or* by linguistic analysis using the style guidelines (Chapter 7) and the expressions to avoid (Appendix 3)". Doing something definite, even mechanical, is — I find — the best way to fight oneself out of the doldrums.

Step 14 See if the first draft needs major alterations

When the first draft has been completed, it must be put away for at least 3 days and then read over in its entirety. Does the whole make logical sense? Are any crucial data missing or uncertain? Probably, some detours, excursions, and divergent lines of reasoning will become evident: are they major or minor, and should they be repaired or eliminated? Almost certainly, convincing transitions from one topic or argument to another will be lacking. With luck, though, the author will feel that with skillful adjustment and a fair amount of vigorous polishing, the arguments and evidence will be seen to march, straight and strong as the columns of a Greek temple, to the conclusions stated in the finale. If so, the author should proceed to Step 15.

Step 15 Have the illustrations prepared in final form

Some illustrations may need to be professionally prepared for presentation to the examiners, in as many copies of the dissertation as required. This should be embarked on in timely fashion.

Step 16 Polish the prose

The advice here is exactly as in Chapter 7, the object always being clean, spare, logical, scientific prose without grammatical errors of the kind most common in science. Deploy the style guidelines (p 47) again if you find them helpful:

1 Write simply and concisely
2 Check the meaning and relationships of every word
3 Seek and release hidden verbs of action
4 Split up noun clusters and stacked modifiers

and mine the books by Lindsay and by Zeiger (p 91) for examples and exercises in addition to those in Appendices 1–4.

Step 17 Rewrite the title, and structure the abstract

Tell students that the title of their dissertation, even if it was assigned at the beginning of the investigations, is rarely immutable and should be thought through again now that the dissertation is almost finished. It is important that the title reflect the content accurately; it is rare for research to pan out just as was originally envisaged. Even if the title is non-negotiable (which is rare), a subtitle can be added that will clarify the content or deflect criticism of any failure to meet expectations.

An abstract may or may not be required by the examiners. However, an abstract or an "executive summary" can never come amiss and should always precede the text, in neatly structured form. An abstract is certainly required by *Dissertation Abstracts International* if the student wishes the dissertation to be listed there, and its form may be specified there too.

Step 18 Request private review by three critics

Desirable as this may be, finding three readers willing to read a whole dissertation will not be easy. The three types of reader to look for, nevertheless, are: someone who is close to the subject, a scientist rather more distant, and a non-scientist. They should be given a deadline, asked to put their comments in writing — not in continuous form but as notes, if that is quicker and easier for them — and they should be thanked before the comments are read. Whether this is also the moment to submit the complete draft to the research supervisor, or whether the author should wait until after the friendly critics' comments have been considered and any consequent changes have been made, is a matter for individual judgment, possibly governed by time constraints.

Step 19 Re-read the Instructions to Authors and make any necessary adjustments

The instructions on dissertations are usually less detailed and exigent than those for a journal article and most will have been absorbed, and adopted, at Step 7. But (for example) the number of copies and mode of binding specified may have changed since Step 7 was completed, or have been forgotten. Warn students to take the same care as in journal articles with respect to citation of private communications and the inclusion of other workers' names in the Acknowledgments section.

Step 20 Revise as many times as necessary

Only one further revision may be necessary (fewer than when co-authors are involved), in conformity with the comments from the selected readers and especially the supervisor.

Step 21 Submit to the awarding body

This will be a university or other examining body. The student must take care to submit the dissertation to the right office and person, in the right number of copies, together with the correct accompanying documents to identify the author, the supervisor, the department, and the date.

Step 22 Analyze the editor's decision letter and respond appropriately

For a dissertation, the decision-maker is not an editor but the external and internal examiners. But in this case there is no opportunity to consider the reactions at leisure; the student will have to answer immediately, whether to friendly or to hostile questioning. For this step, then, substitute the following wording: "Prepare yourself thoroughly, well in advance as well as during the few days before the oral examination, to respond concisely to close, probing questioning by discerning examiners."

The list of steps relevant to the dissertation that your students collectively compile should look something like Table 12.1.

Table 12.1 Writing a good research dissertation

Step 1	Ask yourself whether the time is right
Step 2	Clarify your conclusions by preparing tables and figures complete with titles and footnotes
Step 3	Decide who should be in the Acknowledgments
Step 4	Take ethical matters into consideration
Step 5	In relating your conclusions to the existing body of knowledge, be judiciously selective
Step 6	Write a working title and abstract
Step 7	Study the awarding body's Instructions to Authors
Step 8	Set up section files for both main and subheadings
Step 9	Fill section files with relevant brief notes in any order, to form "ragbags"
Step 10	Arrange the contents of the ragbags logically
Step 11	Finalize the design and content of tables and figures
Step 12	Make a topic outline and consider writing a sentence outline
Step 13	Write the first draft continuously, collecting references as you go
Step 14	See if the first draft needs major alterations
Step 15	Have the illustrations prepared professionally where necessary
Step 16	Polish the prose and clarify the style
Step 17	Rewrite the title, and structure the abstract
Step 18	Seek private review by three critics if possible
Step 19	Re-read the Instructions to Authors and make any necessary adjustments
Step 20	Revise as many times as necessary
Step 21	Submit the dissertation to the examining body
Step 22	Prepare yourself thoroughly to answer examiners' questions concisely

Summary of Chapter 12

Writing a research dissertation

- If considered desirable by the research supervisors in your institution, give a short course on searching the literature and compiling an annotated bibliography to guide and plan research for the dissertation, near the beginning of the graduate students' program.

- The course described in this chapter is, however, best given near the beginning of a graduate student's last year, after the course on writing a journal article.

- Dissuade students from writing a dissertation of excessive length.

- List differences between dissertations and journal articles.

- Take your class through a consideration of the steps in writing a journal article and their applicability to writing a PhD dissertation, to arrive at a table of steps similar to Table 12.1.

13
Research grant proposals

Maureen E Goode* and F Peter Woodford

OBJECTIVES:
To enable mature scientists to write research proposals, using steps similar to those for writing a journal article, that will impress funding bodies.

To provide sources of information from various funding bodies that are available to applicants.

To emphasize the importance of deadlines and the need to start the process long before the neophyte thinks necessary.

YOU SHOULD READ:
L Reif-Lehrer. *Grant application writer's handbook.* Boston (MA): Jones & Bartlett; 1995.

Maeve O'Connor. *Writing successfully in science.* London and New York: Chapman & Hall; 1991.
David Lindsay. *A guide to scientific writing.* 2nd ed. Melbourne: Longman; 1990.
Both these books contain chapters on applying for grants.

YOU WILL NEED:
If your authors are applying to the National Institutes of Health, USA:

Division of Research Grants, National Institutes of Health, Bethesda (MD), 1993:
Preparing a research grant application to the National Institutes of Health: selected articles.
Helpful hints: preparing an NIH research grant application.

Otherwise, the grant application forms and instructions from a funding agency appropriate to your class.

YOU SHOULD CONSULT:
A research supervisor in your institution if you are not yourself acting in this capacity.

* *Department of Scientific Publications, The University of Texas M D Anderson Cancer Center, Houston, Texas, USA*

TIMING:
This is best taught as a single session at the end of the course on writing a journal article, perhaps as a question-and-answer seminar in which the students themselves devise adaptations of the steps of Table 1.1, p 9.

However, if the students have not been following the course on writing a journal article the material can still be presented in a course entirely devoted to grant proposal writing as a series of very similar steps.

STUDENT REQUIREMENTS:
A commitment, by the advisor of each student who enrolls, to read and comment on the mock grant proposal that is the major assignment of the course (see Step 15).

The principles underlying the writing of a journal article apply to writing a research proposal: in both, the objective is a logical structure, clear and succinct phraseology, and flawless reasoning. Consequently, much the same framework recommended for writing a journal article can be used to support the preparation of a research proposal. The steps of Table 1.1 (p 9) appear in Table 13.1, in slightly different order and minimally modified, but with three of the steps relating to the construction of tables and figures omitted because results of the proposed investigation have yet to be obtained.

There are three phases in writing a grant proposal:

- designing a good scientific study,
- writing a clear and convincing research plan, and
- assembling all the other information required by the funding agency, including the technical assistance, equipment and funding requested, and institutional approval.

Table 13.1, based on the steps recommended for writing a journal article, will be seen to fall naturally into these three phases, which we have headed Preparation, Writing the Research Plan, and Completion.

Table 13.1 Writing a successful research grant proposal

	PREPARATION
Step 1	Ask yourself whether the time is right
Step 2	Write a working title and abstract
Step 3	Choose the target funding agency and make notes on its documentation
Step 4	Consider the ethics of applying to different agencies
Step 5	Decide who will be co-authors
	WRITING THE RESEARCH PLAN
Step 6	Relate your proposal to the existing body of knowledge
Step 7	Name the main sections, and open section files bearing those names
Step 8	Fill the section files with relevant brief notes in any order, like a "ragbag"
Step 9	Arrange the items in the ragbags logically
Step 10	Make a topic outline and perhaps a sentence outline
Step 11	Write the first draft continuously, collecting references as you go
Step 12	See if the first draft needs major alterations
Step 13	Polish the prose
Step 14	Rewrite the title, and structure the abstract
Step 15	Request informal, friendly review by three critics and your co-proposers
	COMPLETION
Step 16	Re-read the agency's documents and make any necessary adjustments
Step 17	Revise as many times as necessary
Step 18	Submit the proposal to the agency
Step 19	Analyze the agency's decision letter and respond appropriately

Preparation

Step 1 Ask yourself whether the time is right

The timing of actually submitting a research proposal is governed by external constraints such as when the present funding is due to run out and when the next (or next-but-one) round of consideration of applications is due. But above and beyond this, one must consider

- whether a good case can be made for the *necessity* of research on the question posed,
- whether the applicant is in a position to assemble an appropriate team — which may need to be international for some funding agencies (for instance, the European Commission's Framework Programme) — with the relevant variety of skills and competences to address the question that will be posed, and
- whether the applicant has obtained preliminary results indicating likely success in the proposed investigation.

This last consideration is of great importance in determining the outcome of the application, and is discussed further below.

Preparation of a research grant proposal must start at least 6 months before the deadline for submission. It may take much longer than this if several co-workers in distant departments are concerned — even with e-mail! Discourage students' natural inclination to wait until the most convincing preliminary results possible have been obtained: preparation must start virtually as soon as things look promising, and work on the proposal should proceed while the preliminary results are strengthened.

Step 2 Write a working title and abstract

These will define the purpose and especially the scope of the projected research; the reason for writing them first is to clarify the project in the author's mind, so that a suitable granting agency can be selected.

Step 3 Choose the target funding agency and make notes on its documentation

Applying for funds to an inappropriate agency is a waste of valuable time and may even earn the applicant a bad reputation among the reviewers used by the agencies. An agency whose purpose is to foster women's health would not waste a moment examining a proposal to study prostate cancer, even if it were written by angels. Some granting agencies have misleading or uninformative names (March of Dimes, Rockefeller Foundation), but all of them put out statements announcing the categories in which they award grants to support their stated goals, and a proposal sent to a given agency must fit one or more of those categories if it is to have any chance of succeeding. Sometimes this is a matter of presentation: a proposal to study a rare disease, for example, will stand no chance if presented as if it were a major public health problem, but could well be regarded favorably if presented as a generalizable model for the aging process.

The students must, therefore, having taken advice from colleagues and mentors as to likely agencies, obtain the mission statements and grant application forms from two or three of them and match their own working title and abstract carefully against the statements of purpose and scope to be found there. As they read, students must assess against local circumstances the agency's submission deadlines and any major constraints on applications. Train your students to begin the application process by reading the entire grant application and its instructions, not just the part they think applies to them.

Each country has a different set of possible agencies, ranging from large, broad-scale ones like the National Institutes of Health (NIH) and the National Science Foundation in the USA, the Medical (and other) Research Councils in the UK and in Canada, the Central Scientific and Industrial Research Organization in Australia, and the Wellcome Trust in the UK to smaller, specialized ones like the Macy Foundation in the USA and similar private foundations in many countries. In Western Europe an additional resource is the Framework Programme of the European Commission, which exists to foster transnational research cooperation and to support international trading goals. In what follows, we use the NIH application process as a model that is often followed elsewhere, but it should be remembered that these practices are not universal across the agencies, some of which are even tighter in their requirements while others are more liberal and less oriented toward immediately discernible goals.

Most granting agencies, including the NIH and the European Commission, will advise inquirers as to whether a study is appropriate for the agency. Some make available copies of grant proposals that have been funded; you might obtain one of them for class discussion. And some agencies also disseminate information through their home pages on the Internet. Warn your students to obtain the current, not last year's, set of proposal forms and accompanying documents: conditions often change, and not so subtly either.

Grant proposal writing is, like all forms of communication, highly dependent for success on an awareness of the audience being addressed — in this case the reviewers acting for the agency in a well-defined review process that is usually described in the proposal documents. Every agency uses a group of peer reviewers (scientists with a broad and sometimes detailed knowledge of the area), and some agencies include lay members on the panel. These reviewers usually meet periodically to debate the relative merits of the applications. The NIH calls these their study sections, consisting usually of a dozen reviewers meeting three times a year. But additional external reviewers may be asked to provide specialized views on particular applications, and in bodies such as the European Commission the views of all the scientific experts may be tempered by considerations of "national balance", because one goal of the Commission's research program is to enhance the research capability of countries lacking experienced research teams by linking them with teams in countries with a more developed research tradition. Non-scientists may also be included in the review process. The NIH has issued two short video recordings on writing grant proposals and the peer review process that you could show your class (telephone or fax the NIH Grants Information Office).

You may also be able to invite a colleague who acts as a reviewer for grant applications or is a member of an NIH study section to give a guest lecture about the parts of a grant application reviewers tend to concentrate on. Whether you do or not, you can usefully point out to students that faculty members who fulfill these roles will always be an appropriate source of information about the kind of people who will be considering their grant proposals. It is not unethical to approach these reviewers for information or advice, even in the very field in which the student will be applying, as reviewers are always debarred from making or influencing judgments on applications from their own institution. Some universities have Websites devoted to "grantsmanship".

Step 4 Consider the ethics of applying to different agencies

Similar principles apply here as in submitting an article for publication. In Europe, the same proposal must not be made to two or more agencies simultaneously. It is permissible to apply for different *parts* of an investigation to different agencies, provided the parts can be cleanly separated. In the USA, the same proposal may be submitted to two or more agencies if each is informed and the applicant agrees to accept funding from only one agency.

Step 5 Decide who will be co-authors

When the target agency is chosen, several personnel matters must be clarified: who will be the principal investigator, who will be named as co-applicants, who is willing to be committed to supporting the project by giving technical assistance or providing special services. These matters affect the magnitude of the award requested, and they cannot be sorted out at a later stage without rancor.

In the final application the named co-workers will need to provide a complete curriculum vitae or a biographical sketch, with references to such of their publications that indicate competence or experience in the parts of the investigation for which they will be responsible. They should start compiling this information now.

Suggest that your students obtain from colleagues or mentors one or more successful research grant proposals that can be considered, or assessed, as a model.

Writing the research plan

Step 6 Relate your proposal to the existing body of knowledge

The rationale of the research plan will depend on what is known and has been published already on the subject, where the gaps in knowledge lie and why it is important to fill them, and why the applicant believes these gaps can be filled by the method proposed. As with the introduction to a journal article, the essential writing skill here resides in first thoroughly understanding all previous work, then selecting only what is relevant, and finally encapsulating the line of argument, with adequate referencing, in purposeful and concise prose.

The applicant will eventually need to persuade reviewers that the projected line of investigation is likely to succeed. Promising preliminary results will strengthen the argument, but only if they are highly convincing.

Now is the time, therefore, to depict them in complete detail for critical scrutiny by co-workers and scientific friends. If this process reveals serious gaps, there may be time to fill them; if it uncovers a flaw in the scientific reasoning, the project should be abandoned and a new direction sought. Either way, time and work will not have been wasted. If, however, the preliminary results seem satisfactory, the applicant is ready to proceed.

Step 7 Name the main sections, and open section files bearing those names

Many agencies specify the sections into which the research plan is to be divided and the order in which they should appear; some leave the structure entirely up to the applicant. For illustrative purposes we here adopt the format specified by the NIH, which names the required headings Abstract, Specific Aims, Background and Significance, Preliminary Studies, and Research Design and Methods. Whatever headings are specified, the way to get started is to open files with those headings at the top, into which all the material will be placed.

Step 8 Fill the section files with relevant brief notes in any order, like a "ragbag"

This step is to help writers, however experienced they may be, to collect all the necessary ideas, concepts, arguments, and prior evidence into the right part of the application and examine them for completeness and relevance. Otherwise, a lot of rewriting may be necessary at a later stage because some pertinent fact has been remembered that requires radical re-ordering of the arguments, or some peripheral irrelevant line of reasoning or thought has crept in that distorts the balance of the whole.

Step 9 Arrange the items in the ragbags logically

Once the writer is satisfied that everything relevant has been included and all irrelevancies have been excluded, the bits and pieces in the ragbags are put in logical order. This soon generates a topic outline for each section.

Step 10 Make a topic outline and perhaps a sentence outline

A topic outline helps everyone, especially in a grant proposal, where the structure must be strong and obvious to the busy reviewer. The nouns and phrases of the topic outline may be used as subheadings to lead a rapid reader through the argument. Whether a sentence outline helps all writers (see pp 34–35) is more debatable.

Step 11 Write the first draft continuously, collecting references as you go

Abstract

With the working abstract and the topic outline in hand, the writer has a good overview of how the whole research plan is to look. The abstract will be put into final shape toward the end of the process.

Specific Aims

The writer should now move on to the specific aims, which should start with one or more sentences stating the over-all purpose of the research and its importance to the field, relating both as closely as possible to the goals of the granting agency (whether improved health, protection of the environment, reduction of hospitalization costs, or improved crop yields) as well as to the long-term research goals of the applicant.

In drafting this section the writer should bear in mind the maximum length specified by the instructions. In the final version, this must certainly not be exceeded — and as our goal is always to be succinct, it is to be hoped that the outcome will actually be shorter than the maximum. However, this should not be an overwhelming consideration when the first draft is written: the main object must be getting the ideas down in logical order, and devil take the exact wording, the style, and the length. Subsequent editing (Step 13) will take care of all those matters.

The aims must be stated as hypotheses, and any declarative statements must be justified. Teach your students to ask themselves "How?" and "Why?" after making such a statement; for instance, if a writer states "Pig disease X is the best animal model for human disease Y", he or she must explain why. The reviewer will certainly ask such questions, and if they are not immediately answered the grant may not be awarded.

Advise your students to ask themselves "What makes my project stand out: the novelty of my hypothesis? my unique population, geographical base, or cell lines? my institution's unusual combination of experts?" Emphasizing such special features will make reviewers think more about the project's merits.

On the other hand, the section on specific aims must avoid going into detail on method or approach, which comes in a later section, and indeed the aims should be expressed not in terms of what will be done but in terms of what will be deduced: thus, not "To analyze 100 samples from patients with Z cancer by Southern blotting with the *ras* gene" — which is certainly achievable, but leaves the question "Why?" unanswered — but "To determine whether the *ras* gene is rearranged in Z cancer, by use of Southern blotting of 100 appropriate patient samples". A clue here is to use conclusion-oriented verbs like *determine, discover, elucidate,* not process-oriented verbs like *study, perform, measure.*

There can be as many specific aims in the proposal as there is time to achieve in the period for which the grant is requested. They should not be interdependent, but each capable of being pursued if others fail.

To keep the number and complexity of specific aims within bounds, students should always construct a time line, even if the agency does not require it, showing estimates of when each stage of the study will be completed. These estimates should be thoroughly discussed with supervisors or more experienced researchers, as grants will not be awarded if the estimates are unrealistic.

Large chunks of the rest of the research plan should if possible be written in single sessions of several hours, so as to achieve a sense of unity and logical flow. As with a journal article, when any reference is cited a note should be made, and at the end of each session steps should be taken to ensure that a photocopy of the reference is in the writer's possession by the time that the first draft is put aside prior to revision.

Background and Significance

The importance of selecting the vital facts from all those that could be cited is emphasized earlier, at Step 6. Ruthless editing and rewriting are essential to spare the reviewers unnecessary detail and to convince them of the writer's analytical power. The citation of facts of dubious relevance may make reviewers think that the writer does not quite understand the field and has no discrimination.

Preliminary Studies

The purpose of this section is to convince reviewers that the specific aims can be achieved by the applicant's team. Of course, if the preliminary work has been published, the reference(s) should be cited; if this was in a prestigious journal, its name should be inserted in the text along with the name of the first author ("This work was published in January in *Nature* [Bosh et al. 1997]"). If the principal applicant was the first author on the paper, so much the better; if one of the others, suggest that the applicant use the first person ("My colleagues and I [Bosh et al. 1997] concluded that . . .") so that the reviewer may deduce that the applicant has the track record to complete the proposed studies. Proposal writers should also supply copies of papers cited, especially those that are accepted but still "in press".

Applicants should not present so many preliminary results that the project seems virtually complete and not worth funding. More applicants will probably find themselves in the opposite position, with few relevant results, in which case they must find ways of assuring the reader of this section that they have mastered the relevant techniques and/or obtained the necessary special materials or organisms.

Research Design and Methods

This section describes both the rationale of the proposed experiments or observations and the methods to be used. Both are for the benefit of critical reviewers, to enable them to judge whether the study hypotheses will be adequately addressed and whether the experiments are likely to give usable results. The student must constantly ask "Why are these experiments the best ones to test the hypothesis?"

A flow chart, or decision tree, may be useful to clarify, for the proposal writer in the first instance, how the materials will be apportioned or the experiments tackled in response to the result at each stage. This may also be included in the final version for the reviewers' benefit.

If this section is divided into subsections corresponding to each of the specific aims, caution your students to phrase the latter in exactly the same way each time they are cited.

Determining the degree of detail for the methodology section is a matter of judgment that depends very much on the applicant's awareness of the kind of scientist who is likely to review the application: it is important not to talk down to peer scientists, but equally important not to baffle them with esoteric references to specialized techniques.

There is, however, no doubt of the need for thorough discussion of the rationale and timing for moving from one phase of the investigation to another, with clear statements of what conclusions would be drawn from each possible result. Contingency plans in response to unexpected or unwanted results or unsuccessful approaches should also be made explicit, as well as the logic of choosing one line of approach over another. Thus, not "We will isolate the protein by method A, and if this fails by method B", but "We will first try to isolate the protein by method A, which is fast and cheap, but if this approach fails because the protein is present in such small quantity we will use method B, which is more sensitive but takes longer and is more expensive."

The completed first draft should be put away while all the papers that have been cited — published and unpublished — are carefully read over. The writer must be quite sure that they have not been misquoted through a failure of memory (remembering, always, that one of the original authors may be among the reviewers of the application!).

Step 12 See if the first draft needs major alterations

Now, having read all the papers cited, the student pulls the first draft out of the drawer or computer and examines it for structural defects, failure of logic, and misquotation. Any deficiencies must be rectified before the actual wording is worked on.

Step 13 Polish the prose

When the structure of the various sections, and of the whole proposal, is rock-solid, students can begin to hone the wording so as to make it as clear, succinct, vigorous, and direct as possible. All the advice in Chapter 7 applies.

Step 14 Rewrite the title, and structure the abstract

A reviewer may never get beyond the title and abstract if they are not arresting, persuasive, and in line with the goals of the granting agency. This step is therefore of vital importance.

Step 15 Request informal, friendly review by three critics and your co-proposers

The draft will now be in a fit state to be subjected to critical scrutiny by everyone who is mentioned as a member of the research team, but also by anyone else with experience of applying for grants who is in a similar or related field and who can be cajoled into considering the draft seriously and making constructive suggestions. As with a journal article, the following important rules apply: Ask for the comments in writing, say to which agency you are applying, and supply the whole text along with illustrations and references; thank the commentator for the suggestions immediately upon receipt, before reading them; take them away and consider them quietly; take any corrective action only after careful thought.

As a graduate student's most appropriate critic is his or her advisor, you might make it a condition of enrolling in this class that each student gets his or her advisor to be committed to read and criticize the student's (mock) grant proposal (see the head of this chapter, p 103). Most will agree to such a commitment when they realize that they will probably benefit from the exercise themselves.

Having students criticize one another's draft proposals is a possibility at this stage, but what is required most here is the eye of experience, and the experiment is not recommended unless about half of your class members have such experience.

Parts of the grant proposal other than the research plan, namely all the financial aspects — cost of hiring technicians, length of contract, cost of equipment and materials requested with any alternatives or possibilities of sharing, percentage of overheads demanded locally versus that allowed by the granting agency — will need to be reviewed by the head(s) of department(s) and administrators of the research institution. The need for signed endorsements by these people will be specified in the proposal documentation. Obtaining these signatures is best left until after the research plan and associated expenses have been agreed in principle by the proposer's colleagues and line manager, but preliminary inquiries as to the administrators' availability are in order well before then.

Completion

Step 16 Re-read the agency's documents and make any necessary adjustments

While the co-workers are considering the polished draft, the principal writer should read the whole of the documentation again, making sure that no sin of omission or commission will result in the proposal's being turned back for adjustment even before it is sent to reviewers.

Step 17 Revise as many times as necessary

The larger the number of members of the research team and the more complex the departmental or institutional or inter-institutional arrangements, the more revisions and further circulation of drafts are likely to be needed. This is why the whole process should be started many months before the innocent first-time proposal writer imagines it to be necessary.

Step 18 Submit the proposal to the agency

If Step 16 has been conscientiously followed, the submission of the final version to the agency should be easy. Nevertheless, advise students

- never to leave it until the last possible day before the deadline, for strikes of various kinds, including by the mail service, can instantly undo many months of work
- to send as many copies as the agency requires
- to register the packet so as to ensure its safe arrival at the right destination
- to enclose a self-addressed and stamped card acknowledging receipt by the agency, whether or not this is requested
- to keep an exact copy of everything that is sent.

Step 19 Analyze the agency's decision letter and respond appropriately

Weeks later, the dreaded decision letter will arrive. It is usually futile to dispute negative responses and best to accept whatever is offered, even if it is less than requested. If the criticisms are minor and reconsideration or even conditional acceptance is offered, the comments should be politely countered or gratefully accepted and the proposal returned with the modifications clearly identified. But if the response is a straight refusal, the research plan should be radically revised or replaced by a new one, and a different agency should be approached. Granting agencies are for the most part so inundated with requests that they have little time or patience to consider reconciling fundamentally opposing views.

Summary of Chapter 13

Writing a research grant proposal

- Teach this subject to advanced graduate students and/or more mature scientists either as an add-on session to the journal article course or as a complete course in its own right.

- Emphasize that the process must start *at least* 6 months before the submission deadline.

- Apply the same principles as for a journal article, stressing the importance of thorough preparation, clarification of purpose and scope through the writing of a working title and abstract, and well-informed study of the characteristics of reviewers before any actual writing of the text is begun.

- Point out sources of information on the review process used by different grant agencies, including (a) the agencies' documentation and video recordings and (b) those colleagues of yours who act as reviewers of grant applications.

- Fit a series of recommended steps, closely similar to the steps in writing a journal article (Chapter 1), into a tripartite structure:
  Preparation
  Writing the Research Plan
  Completion

- Use material from Chapter 7 to aid students in editing their text.

- Stress that time is of the essence, and that the task does not end with completion of the research plan.

14
Communicating science to the lay public

F Peter Woodford and Barbara Gastel

OBJECTIVES:
To point out stark differences from writing a journal article or delivering a scientific talk.

To provide guidance in writing a book, preparing a magazine article, giving a talk, or being interviewed by radio, television, or newspaper journalists — all for non-scientific audiences.

TIMING:
One to four sessions can be given as a stand-alone course or at the end of a course on writing a journal article or giving a scientific talk, at virtually any stage of a scientist's career after the first year of graduate school. The number of sessions will depend on the depth with which you want to treat the different kinds of writing for the non-scientist, the different kinds of talk for a lay audience, and the handling of the various forms of the popular media.

YOU MIGHT CONSULT:
M Shortland and J Gregory. *Communicating science: a handbook.* Harlow, Essex: Longman; 1991 (reprinted 1996).

Barbara Gastel. *Presenting science to the public.* Philadelphia (PA): ISI Press; 1983.

Barbara Gastel. *Health writer's handbook.* Ames (IA): Iowa State University Press; 1998.

D Blum and M Knudson, editors. *A field guide for science writers.* New York: Oxford University Press; 1997.

Communicating science news: a guide for public information officers, scientists and physicians. Greenlawn (NY): National Association of Science Writers (NASW); 1996. Available online at http://www.nasw.org.

YOU WILL NEED:
Copies of *Science, New Scientist,* possibly *American Scientist;* any

well-written popular science book such as Dava Sobel's *Longitude* (New York: Walker; 1995; London: Fourth Estates; 1996); video recordings of any recent TV program on a scientific subject, for instance from the BBC *Horizon* series, both well and badly executed.

Although few scientists pursue careers in popular science communication, all graduate students and mature scientists can benefit from learning some basic facts in this realm.

Why communicate with the lay public?

Run lightly over the reasons, in case your students do not provide these answers for themselves:

- If research is publicly funded, the public deserves to be informed about it.
- Informing the public about science can help promote science and scientific institutions and encourage people to seek scientific careers.
- Conveying science news into the public domain can produce cross-fertilization between fields of science — because scientists from other fields form part of the "lay audience" for your field.
- Members of the public need scientific and technical information they can understand, if they are to make sound political decisions.

Ways of communicating with lay persons, I: Presenting science to the public oneself

Speaking to lay groups

This is the opportunity most likely to come to your students. Many lay groups, ranging from elementary school classes to special-interest groups, the Rotary and Lions Clubs, and retirees' organizations, are interested in talks on scientific topics. Invitations to give such a talk may come at quite an early stage in a scientist's career — for instance, if he or she has discovered something potentially lucrative in the raw material of a local industry and this has been reported in a local newspaper or trade journal. Discuss making oneself available to give such talks, and note that some institutions and professional organizations maintain lists of speakers and topics on which they are prepared to speak.

Later in their career, scientists may well find themselves having to talk to a largely non-scientific board of directors to persuade them to buy some expensive piece of equipment or massively expand the research program, and experience gained in these less formal settings should stand them in good stead.

Writing magazine articles

Show examples of such articles in *Science*, *New Scientist*, and the like, stressing that although most are written by professional science writers, such magazines will always consider unsolicited, topical articles from others, particularly if they are unusual and off-beat. Initial contact with the magazines can usefully be made through the correspondence columns, from which the student can gauge the sort of thing the magazine appreciates. These magazines sometimes run competitions with attractive prizes for the best article intelligible to a lay audience, either on a stated topic or on one chosen by the competitor, and sometimes there is a special prize for young entrants.[1] Identify some potential sites of publication and discuss how to contact editors, including those of campus publications. If you teach at a university with a journalism department, mention (with their prior permission!) faculty members who can offer guidance.

Writing science books for the public

This more ambitious goal usually comes only after a scientist's career is established. But it is worth any scientist's while to study the techniques used in a successful book of this kind, for instance the use of personal reminiscence, quotation from famous people (inspiration may be sought in anthologies like that cited in the Bibliography, p 191), striking analogy with everyday objects or processes, sparkling metaphor, and short, bright sentences. These can do no harm in any kind of communication.

Examples abound, but look at a few in a recent book in the history of science (classically regarded as a surefire soporific) that has taken the public by storm, *Longitude*. After the author, Dava Sobel, has elegantly clarified how important the accurate determination of longitude was to navigators the world over, the loss of life and goods to which the failure to do so inevitably led, and the inadequacy of the astronomical approach much favored by monarchs, she introduces her hero.

> English clockmaker John Harrison, a mechanical genius who pioneered the science of portable precision timekeeping, devoted his life to this quest. He accomplished what Newton had feared was impossible: he invented a clock that would carry the true time from the home port, like an eternal flame, to any remote corner of the world.

> Harrison, a man of simple birth and high intelligence, crossed swords with the leading lights of his day. He made a special enemy of the Reverend Nevil Maskelyne, the fifth astronomer royal . . . .
>
> An aged, exhausted Harrison, taken under the wing of King George III, ultimately claimed his rightful monetary reward in 1773 — after forty struggling years of political intrigue, international warfare, academic backbiting, scientific revolution, and economic upheaval.

With such an introduction, what reader could fail to turn eagerly to the next page?

The story your students have to tell may be less dramatic than Sobel's (though who, before this, would have thought the search for a solution to the longitude problem anything but dull?), but an audience's interest can always be heightened by an introduction in well-chosen, lively words, especially if there is a hint of the rivalry and tension that often underlie the pursuit of scientific truth. Take this piece, introducing the projected use of tiny microbiological agents instead of brute chemicals to transform cheap steroids into costly, valuable corticosteroid drugs. (Ask your students to forgive the gender-insensitive language; the example is from an earlier era.)

> The organic chemist, backed by all the resources of a modern laboratory, and by all the knowledge acquired in more than a century's tremendous activity, is still confronted at times by a serious limitation of his craft. Although he may point to a single atom in a bewilderingly complex formula and announce with complete confidence that he is going to replace *that* hydrogen by a bromine atom in one step, his admiring but never completely credulous lay friends may not be surprised to learn that there lurks within the same structure another atom, of no markedly different aspect to the untutored eye, which he is incapable of altering without going a great deal out of his way.

There is no metaphor or simile here, but a distinct sense of struggle (the lone scientist confronting the perverse obduracy of Nature) and mild distrust (the scientist versus his lay friends) with which the readers or auditors can identify. And underlying the statement of an apparently intractable problem there is a confident sense that the writer is going to produce a solution, or at least be eloquent in proposing one.

Communicating through the electronic media

Apart from local television and radio stations, there are now opportunities to communicate directly with the unseen public via the World Wide Web. You may be able to direct students to people on campus who can offer guidance or collaboration. This route may be the one with the greatest feedback potential by which authors may judge their degree of success in stimulating interest.

Basic principles for any of the above
Give your students three questions to use as touchstones.

"For whom am I writing or speaking?"
This is the most important consideration — to be vividly aware, especially for talks, of members of the intended audience and their characteristics: age range and hearing ability; distances they will have had to travel to attend; whether they are likely to be tired because of the time of day; ethnic background; first language; acquaintance with science, if any; subjects of any recent talks to this group; and of course the special interests of the group. The reader of a newspaper or magazine can always stop reading if the article is outside his or her interests, but a seated audience usually finds it embarrassing to leave an uninteresting, or at least *undirected,* talk.

Whether for a talk or an article, authors must drum up interest *for this particular audience* in any way they can think of — for instance, use of:

- a **catchy title**, coupled with an informative subtitle if the title is so catchy that it is meaningless without explanation. For instance, the talk on a microbiological subject mentioned on p 121 might be entitled
  Living Reagents,
  or How to make wonder drugs cheaply without using chemicals
- a lively, surprising, or even shocking **"display" sentence** selected from the prepared text and either printed in bold type above the piece or projected as the first slide, for example:
  Raw sewage provides the most promising source of today's technology
- **condensation** of the essence of the article's message into the first few sentences, so as to provide an overview before plunging into detail, for example:
  I'm going to tell you today about a recently discovered solution to an age-old problem: how do fats, which are not soluble in water, travel around in the blood, the bile, and all the other watery liquids that make up most of our body weight? Ah-ha, you will say, we know the answer to that: they are wrapped up in blankets of water-soluble proteins. Well, besides that, there are much smaller particles in which they can travel. These have been given the name *micelles.*
- **simple, vigorous language** using strong verbs and few abstractions. Note, in the quotation above, how in opening a talk the author has avoided the impersonal (not *This talk will be about,* but *I'm going to tell you*), and used simple terms wherever possible without distorting the sense (not *lipids*

but *fats*; not *diffuse* or *are water-borne* but *travel around*; not *aqueous* but *watery*). In addition, the author refers to a well-known fact the audience can relate to (the high proportion of water in the body); uses a homely metaphor, *blankets*, for lipoproteins, simultaneously hinting that the audience knows about them but informing those who didn't, without being condescending; and ends with a surprise twist adding to that knowledge without negating or devaluing it, and introducing a new term that is the subject of the talk, which virtually no one in the audience will have heard of. A catchy title for this talk might be *Small is Beautiful, or How fats get around the body.*

- plenty of **metaphors and similes** that this audience can relate to or identify with. Once the author gets into this frame of mind and embarks on this train of thought, the similes will readily spring to mind. For micelles, for instance, the activities of boating and swimming would provide ample scope for metaphor. Let your students' imagination fly, and the similes will burst out like confetti — look at "like an eternal flame" in the passage from *Longitude*, sparked off by the verb *carry* and the thought of a single precious thing transported from place to place.
- our old friends, **brevity and succinctness**. No need to elaborate further on these qualities, whose virtues have been well and truly sung elsewhere in this book, except to say that the cautious qualifications and hedging appropriate to a scientific audience is far from appropriate with this one — they only obscure the message and impede comprehension, which may be difficult enough for an audience grappling with unfamiliar concepts.

But before fixing these generalities in the front of their mind, authors must ask themselves the following two further questions.

"What, exactly, do I want to convey, and why should it interest this audience?"

There should be at most three messages per talk, for there must be time to lead up to and explain each one clearly. Speakers should decide which is their most important message and make it stand out unmistakably from the rest, making sure that the audience will consider it important *to them* by relating it to the audience's suspected or known interests. These will always include money; their own or their children's health or sex life; the problems of old age and the environment; as well as anything more specific that can be found out about them. After this, the task of making the talk or article interesting and entertaining as well as important should be child's play.

"Why now?"
Topicality always interests an audience, so speakers should take care to refer, explicitly, to the topicality of the subject of their article or talk. If their subject is a recent discovery or breakthrough by the author or by colleagues, so much the better; but it need not be so personal. Why not, for instance, the reasons why the latest Nobel prize was awarded? The latest exciting idea or project in biomedicine, biotechnology, or plant ecology? A centenary or anniversary of a discovery? Even something not recent at all can be made topical by explaining why the subject has attracted attention at the local scientific institution for the past 20 years, and what stage the research has now reached.

Planning and delivery of a talk to lay persons

Chapter 11 goes into detail about how to plan, prepare, and deliver a scientific talk to a scientific audience. Much of this advice is pertinent for a non-scientific audience too, as long as the speaker bears the above-mentioned differences in mind. The slides will have to be quite different, and the language should be simplified but not patronizing, free of technical jargon, obscure abbreviations, and tortuous qualifications. Speakers should go carefully through their prepared talk looking for terms that need to be either weeded out or clearly explained. They will not go far wrong if they think constantly and hard about the audience, its degree of general education, knowledge, comprehension, and sophistication. Jokes and humorous anecdotes may play a larger part than with a scientific audience, but humor must be circumspectly deployed and strictly relevant. Sources like the one given in the bibliography, p 191, may yield suitable gems.

Encourage your students to give talks of this kind as often as opportunity affords if they wish to be good communicators. Doing so will inevitably help with communication of all kinds, including communication with other scientists.

Ways of communicating with lay persons, II: Working with members of the media

You will need to convey some general points first:

- The word "media" is plural, and it includes local and national newspapers, radio, and television, which vary greatly in content and tone. Scientists should gear what they say to the specific publication or broadcast, and should not judge all media by the performance of some.

- Many scientists see the function of the media as educating the public or promoting science, whereas reporters and presenters see it as mainly to excite interest and even to entertain. Recognizing and acknowledging this difference will save a lot of heartache.
- The popular media operate under tight constraints of time and space; stories in them cannot be as complete as in scientific journals. If they were, few members of the public would want to read, view, or listen to them.
- Some scientific journals decline to publish scientific papers if the research reported in them has been described in detail in popular media. This point will affect students who are in fields in which the major journals have this policy. Where it applies, scientists have to exercise fine judgment on the degree of detail they release to the press.
- Many universities and other research institutions have public information offices ("press offices") to help inform the public about science, which can give professional advice to scientists in the institution as well as directing reporters to them. Advise your students not only to keep that office informed about the progress of their work and its possible significance to a wider public, but also to make a general practice of consulting the press office about any contact with the media.

Advice on being interviewed

Most communication with the popular media occurs through interviews with reporters from newspapers, radio, or television. There is an art in dealing with reporters; offer your students the following pieces of advice.

- *Find out what the reporter is seeking, and consider whether you are well qualified to respond.* The reporter may have obtained your name in error. If you are not well qualified to respond, decline the interview, if possible referring the reporter to the institution's press office, advising that office who else in the institution might be approached.
- Especially if you are inexperienced, *consider inviting someone from the press office to be present* during the interview, or perhaps have such a person coach you beforehand.
- *Ascertain the scientific background of the reporter and the publication represented.* Some reporters have no scientific knowledge, others have advanced degrees in one of the many scientific fields. Their publications range from *Science* and *Science News* to supermarket tabloids and other

"freebies". If the publication seems shaky, consider declining the interview. When you accept an interview, gear your remarks to the reporter's background.

- *Draw on the available guidance in working with the media,* including the local press office if there is one, publications listed at the head of this chapter, and some professional scientific societies.
- *Work within the reporter's constraints.* Reporters for daily media have a very short time to research a story. Find out the reporter's deadline, and respond accordingly. Make sure you take the reporter's name and telephone and fax numbers for later urgent contact.
- *Respect reporters and their audiences.* Reporters and their readers may be ignorant of your field, but do not assume they are stupid, and do not condescend. Watering things down tends to produce indigestible chunks of information in an insipid broth. Aim, rather, at building bridges between the audience's previous knowledge and the new material you are contributing.
- *Try to present material in a form suitable for direct presentation to the audience.* The less "translation" the reporter must do, the more likely it is that your information will be accurately conveyed. If you have visual materials or ideas for pictures, let the reporter know.
- *Be concise and focused.* Concentrate on what is most important. Reporters who have little science background may not know what crucial questions to ask. Take the initiative to present key points if the reporter fails to ask for them. For broadcast reporters, try to provide some "sound bites". But don't say anything, at any stage in the interview, you would not want to see in print or hear on the air. Even if you specify that some words are "off the record", the reporter may forget or ignore your request.
- *Provide some material in writing.* Doing so assists reporters and promotes accuracy. Depending on the reporter's background and your own constraints, the materials can include news releases, journal articles or editorials, or simply lists of key points, preferably in sentence form. Fax machines and e-mail allow you to provide such materials rapidly to reporters almost anywhere.
- *Check reporters' understanding by having them repeat key points in their own words.* A tactful approach is "I'm not sure I've explained that clearly. Won't you summarize it in your own words, so I can clarify if necessary?"

- *Offer to answer questions later.* Make clear that you don't object to being telephoned at home or otherwise contacted subsequently, so the reporter can ask any questions that occur when the story is actually being written.
- *Offer, if time permits, to review a draft for technical accuracy. However, do not demand that the piece receive your approval before publication.* Such a demand runs counter to journalistic practice, and unless the person interviewing you is on the public information staff of your institution will almost certainly be rejected. Limit your suggestions to technical content, and present them with tact.
- *If work you have done may generate considerable media attention, plan accordingly.* Work with the press office of your institution to prepare a news release that will appeal to the media, and then make yourself available for interviews.
- *Be aware of basic differences between media types.* Daily newspaper reporters have only hours to prepare a story. Writers for weeklies or magazines have more time, which makes many scientists prefer them; but remember, you may then be only one of many sources interviewed, so do not be surprised if you are quoted only briefly or not even mentioned by name. TV news stories are very short and rely heavily on visuals, as do TV features, which are longer, but even makers of feature films like to break up lengthy interviews into short snatches (as any video recording will demonstrate). Be ready to present your message(s) in a few telling words, and help to provide suitable images. Radio stories may be enhanced by "natural sound"; try to identify sounds associated with your work.
- *After publication, evaluate your communication with the reporter.* Even the best science reporting will not be as precise, detailed, and cautious as a journal article. Be prepared to swallow your indignation at any failings, and count the blessings the piece may nevertheless bring. Assess how you might communicate more clearly or informatively, and possibly less frankly, in the future.
- *Provide feedback, both favorable and unfavorable.* If a reporter has done an exceptionally fine job, let him or her know: such feedback not only is welcome but also helps to promote skillful science reporting. If the story contains a serious error (which may have been introduced either by the reporter or by an editor), again inform the reporter. Do so in a positive way, noting at the same time the strengths of the piece. Good reporters value feedback that can help avoid repetition of mistakes.

Courses and fellowships in popular science communication

One or two of your students may become particularly interested in popular science communication and may enjoy and benefit from courses and fellowships in this field. Various universities in the USA and elsewhere offer courses in science journalism. Through initiatives such as the Mass Media Science Fellows Program of the American Association for the Advancement of Science (AAAS), graduate students can work at media sites for a summer. Sources of further information are the (US) National Association of Science Writers (http://www.nasw.org) and similar organizations in other countries.

Careers for scientists in popular science communication

These constitute one of the many professional options for trained scientists. Such a career can encompass work as one or more of the following:

- a reporter, writer, or editor for the popular media
- a public information professional at a university, research institution, government agency, scientific society, company, or other institution
- a staff member at a science museum, book publishing company, or other entity involved in popular science communication
- a freelance science writer.

Reference

1 Zetie K. How to write a winner. *Physics World* 1997;10:49. [by the winner of the 1995 Science in Print competition]

Summary of Chapter 14

Communicating science to the lay public

- Discuss the reasons for communicating science to the lay public.
- Present ways for scientists to do this for themselves:
    - speaking to lay groups (an opportunity which may occur for graduate students)
    - writing magazine articles (probably later in their career)
    - writing books on science for the public (certainly later)
    - communicating through the electronic media (at any stage).
- Questions to ask in doing any of the above:
    - For whom am I writing or speaking? How do I capture the attention of different audiences?
    - What message do I want to convey, and why would it be important to this audience?
    - Why now? What is the topical angle?
- Use the material in Chapter 11 to advise students how to plan and deliver a talk to non-scientists, but point out that slides will have to be re-designed, language will have to be free of unexplained technical terms, anecdotes and similes should be richer but strictly relevant, and the whole must be matched to the audience's known and assumed interests.
- Present ways of working with members of the popular media, noting that:
    - the media comprise newspapers, magazines, radio, television
    - scientists and reporters have different perceptions of the functions of the media
    - scientists must recognize reporters' time constraints
    - some journals find prior publication in the media reprehensible
    - the expertise of the institution's public information office is there to be used.
- Give detailed guidance on being interviewed by reporters.
- Note the existence of courses, fellowships, and careers in popular science communication.

Appendix 1

Assignments on style guidelines

Assignment on Style Guideline 1 "Write simply and concisely"

Make these phrases or sentences simple and concise.

At the termination of the experiment . . .
X has the capability of . . .
At a high speed level
The concentration was measured at pre and post of 4 sessions.
S produced an inhibitory effect on the formation of W.
Considerable data has now been accumulated to demonstrate that . . .
Unlike human coronary arteries, which are usually severely diseased in chronic diabetes or after percutaneous transluminal coronary angioplasty, the blood pressure was high.
X formed Y at least an order of magnitude faster when . . .
This result would seem to indicate the possible presence of . . .
Effectiveness of the oral inoculum in producing caries varies widely with the strain of rat; in some cases, rats may become highly caries active, whereas in other strains, the oral inoculum has much less adverse influence.
It was possible to obtain semipreparative (100-μg) quantities of substance X.
Computations were conducted . . .
Cells were observed in a state of protrusion through the membrane.
Subsequently to their entry into the cell
Occupies a juxta-nuclear position
Solvents were pre-cooled at 0°C prior to use.
One lot contained particles greater than 74 μm, and this material was shaken on a sieve prior to use to remove particles in excess of this size.
Figs. 1–3 are photographs of thin-layer chromatograms developed in the solvent system described and are typical of the separations achieved with this chromatographic method.
In the immediate post-bombing period . . .

Assignment on Style Guideline 2 "Check the meaning and relationships of every word"

Analyze these phrases or sentences using Style Guideline 2 and rewrite them using precise wording.

Glyceryl ethers of varying degrees of unsaturation . . .
This method for delipidation can be used on human urine that is superior to that now in use.
The composition of the lymph of fasted cows are unlike depot fat.
Addition of normal serum decreased the number of cells and was therefore not included in the incubation medium.
Microheterogeneity of oligosaccharide structures exist . . .
You might disagree strongly with the tenant of this article.
. . . either from experience or here-say
If the overall consequences of a decision our beneficial then . . .
The problem of transgenic expression of non-transfectable genes . . .
Some workers have reported more *trans*- than *cis*-fatty acids in coconut oil.
Reference 7 reviews this problem.
Based on their respective weights, the mice were placed in cages of varying size.
Following the incubation, the remaining fluid was poured off and the slices washed.
The pollen was extracted, followed by centrifugation, and the upper phase withdrawn.
Transgenic callus and shoots were screened for green fluorescent protein expression using an inverted fluorescence microscope.
Petals were homogenized in a blender, extracted with phenol, and precipitated overnight in ethanol.
Fasting blood was drawn.

Which of these is correct? [requires an appreciation of the meaning of the verb **to approximate**]

1 In the steady state the daily fecal excretion of neutral plus acidic steroids of endogenous origin should approximate the daily synthesis of cholesterol.
2 When radioactive cholesterol is given to patients with every meal, the specific activity of biliary bile acids approximates that of plasma cholesterol after some days.

Assignment on Style Guideline 3 "Seek and release hidden verbs of action"

Correct these examples of pomposity by releasing the hidden verbs. The "nothing verbs" have been highlighted, to help you.

The separation of A from B was **effected** as follows.
Isolation of the tertiary component was **accomplished** and its identification **achieved** by the following manipulation.
Promotion of gene expression was **brought about** by . . .
Transfection of the virus was **achieved** . . .
Ionizing radiation was **applied** to the cells.
Inoculation was **administered** to patients.
Statistical analysis was **performed** on the results.
Conclusions were **produced** based on the experimental outcomes.
Pre-incubation was required before homogenization of the contents of the tubes was **done**.
Hydriodic acid attack on unsaturated ethers **proceeds** at double bonds.
Primary and secondary particle separation was **obtained** by performing electrophoresis.
Injection of the protein was more difficult of **achievement** in older animals due to the frequency of **occurrence** of thrombosis.
Preferential release of monoenoic acids is also **accomplished** in man.
The paper lost its integrity.
Formation of fatty globules was **experienced**.
Frequent vomiting **occurred**.
Several injections were **carried out** following this occurrence.
The following modifications in the methodology have been **implemented**.

Assignment on Style Guideline 4 "Split up noun clusters and stacked modifiers"

Try to disentangle the following stacked modifiers (or "noun clusters") into phrases that make sense.

nonglucose light experiments

measurement of urine free cortisol excretion test

venous occlusive plethysmographic measurements

forearm arterial compliance determination

skin water accumulation

glutathione reductase inhibitor insensitive procedure

green fluorescent protein labeled transgenic tobacco plants

perennial plant leaf surface landscape scans

cellulose acetate electrophoresis procedure

The low density lipoprotein cholesterol-lowering effect of exchanging dietary protein for carbohydrate in normolipidemic human subjects can be involved in the treatment.

New classes of drugs such as angiotensin II receptor blockers will soon be tried to effect smooth muscle cell proliferation inhibition.

The up to now limited availability of . . .

Using the rabbit aorta carotid artery balloon injury, the effect of estrogen in preventing myointimal hyperplasia was studied.

An Italian Parmesan style cheese making facility is to be set up using the traditional labour intensive Wensleydale semi soft methodology.

Assignments on all four style guidelines

Computation of the data have been conducted by means of varied statistical packages.

Adjustment of the electrical circuits required sensitivity of manipulation.

Much recalculation of population frequencies were involved.

Completion of construction of the animal houses was attained after the administration had been involved.

Obese essential hypertensive patients were compared with controls, and the insulin secretion did not change (3.16 vs 3.23 U/h).

CRH was isolated and characterized from sheep hypothalami in 1981 to be a peptide of 41 amino acids.

There was a marked decrease in the incidence of the allele among healthy elderly individuals as compared to young persons.

Patients on estrogen therapy will fail to demonstrate . . .

The frequency of the gene in this group was 6.8 as compared with 5.4 in the general population.

PLC-treated rabbits exhibited a reduction of the number of proliferating plaque cells as compared to control rabbits.

With the introduction of low-dose cyclosporin therapy in relatively healthy individuals such as those with psoriasis the risk:benefit ratio is greatly altered compared to patients receiving transplants.

The former was older in age as compared to the latter.

Based on the observation of a negative correlation between X and Y, we wondered whether . . .

Older subjects had somewhat higher serum X than younger subjects, although males and females of the same age range had similar values.

In a random population, however, men had a higher percentage of abnormal values than women.

In comparison to the UV induced skin lesions, the size of the scalp lesions was smaller.

Unlike human varicose veins, which are usually severely diseased in chronic diabetes or after percutaneous transluminal coronary angioplasty, most animal models employ normal veins.

Proliferation began in the media between 24 to 48 h after injury. Following this period, the smooth muscle cells migrate from the media to the intima and continue to proliferate. The accumulation of smooth muscle cells was at a maximum at 2 weeks in the rat carotid artery model. However, the intima continues thickening in that extracellular matrices are formed and the entire thickness of the intima reaches a maximum at 28 days.

Appendix 2
Editing assignments

This appendix contains some examples of assignments you can give in which students develop editing skills to be applied to their own first drafts by editing (a) tables with their associated wording, (b) abstracts, and (c) whole articles that have been badly constructed and are full of faults of composition.

The articles have many faults, ranging from inept titles to a mixture of formats in the articles' reference lists. The last one (p 156) lacks an abstract. Advise students to make an abstract after reading the whole, stating what they consider the article to be about before they embark on editing it. This may suggest more radical measures than merely correcting the grammar and style.

Editing assignment 1
Poor correlation between text and table

Table A Storage of Brobdingnaldehyde

	Days		
Storage conditions	2	7	14
	% remaining		
Corked container (23°C)	100	43	7
Under nitrogen (0°C)	100	100	98
In vacuo (23°C)	100	100	100

Table 1 shows that brobdingnaldehyde stored under vacuum at room temperature showed no decomposition in 2 weeks. The sample stored under nitrogen with refrigeration also showed very little alteration, but less than 10% of the sample stored in air at room temperature remained after 2 weeks.

FAULTS: The different storage conditions are referred to in the reverse order from that in the table. Apart from being confusing, this brings the conclusion to the wrong place in the text. The other variants are:

TEXT	TABLE
under vacuum	in vacuo
room temperature	23°C
2 weeks	14 days
with refrigeration	at 0°C
less than 10%	[only] 7%
[implied] in air	corked container

Editing assignment 2
Table with a poor title, unnecessary repetition, and illogical grouping

Table X Carbohydrate composition of isolated lilliputamine hexosides*

Swift's Disease	Glucose	Galactose	Galactosamine
Lilliputamine monohexoside	1.00	0.54	–
Lilliputamine dihexoside	1.00	1.07	–
Lilliputamine trihexoside	1.00	0.96	1.17
Lilliputamine tetrahexoside	1.00	1.83	1.19
Gulliver's Disease			
Lilliputamine monohexoside	1.00	0.78	–
Lilliputamine dihexoside	1.00	1.03	–
Lilliputamine trihexoside	1.00	1.24	1.18
Lilliputamine tetrahexoside	1.00	2.03	1.07

* Expressed as the molar ratio with glucose as 1.00.

Improved version

Table Y Carbohydrate composition of lilliputamine hexosides isolated from human spleen

	Molar proportion (glucose = 1.00)	
Lilliputamine	Galactose	Galactosamine
Swift's disease		
Monohexoside	0.54	–*
Dihexoside	1.07	–*
Trihexoside	0.96	1.17
Tetrahexoside	1.83	1.19
Gulliver's disease		
Monohexoside	0.78	–*
Dihexoside	1.03	–*
Trihexoside	1.24	1.18
Tetrahexoside	2.03	1.07

* Not detectable (ratio therefore <0.05).

Further improved

Table Z Similarity of lilliputamine hexosides from spleen in Swift's and Gulliver's diseases

Lilliput-amine hexoside	Swift's disease	Gulliver's disease
	molar proportion[a]	
	Galactose	
Mono-	0.54	0.78
Di-	1.07	1.03
Tri-	0.96	1.24
Tetra-	1.83	2.03
	Galactosamine	
Mono-	–[b]	–[b]
Di-	–[b]	–[b]
Tri-	1.17	1.18
Tetra-	1.19	1.07

[a] Relative to glucose = 1.00.
[b] Not detectable (ratio therefore <0.05).

COMMENT: Once the repetition has been stripped away it becomes more obvious that the important comparison is not between galactose and galactosamine but between the two diseases. When these figures are put side by side, the conclusion leaps out: the galactosamine proportions are similar in the two diseases, the galactose compositions are not.

The meaning of dashes in a table must always be explained (measurement not done, measurement below the limit of detection, or observation not achievable for a reason given).

Editing assignment 3
A table of advanced horror status

Table as submitted

Table 4 Hypolipidemic effects of CGS 26214 in cynomolgus monkeys

Dose Parameter Pre-Versus Post-Treatment

(ug/kg) LDL (Mean ± SEM)

1 (mg/dL) 95±12 74±14

Mean % Change

23**

30 Lp(a) Range

4.5-25 1.4-12.3

Mean % Change

43**

The horrific table improved

(it's now more intelligible, and also more informative)

Table 4 Hypolipidemic effects of CGS 26214 (one daily oral dose for 5 days) in cynomolgus monkeys

Dose (μg/kg)	Lipoprotein measured	Before	After
		mg/dL	
1	LDL	95±12	74±14[a]
	Mean % change	−23[b]	
30	Lp(a)	4.5–25[c]	1.4–12.3[c]
	Mean % change	−43[b]	

[a]Mean ± SEM.
[b]$P < 0.05$.
[c]Range.

COMMENT: Whether a percentage change can really be calculated from two such wide ranges of values is questionable, but tightening up the table in this way brings such doubt to the fore.

Editing assignment 4
A pompous abstract in botany, part 1

MAPPING LEAF SURFACE LANDSCAPES[1]

ABSTRACT Leaf surfaces provide the ecologically relevant landscapes to those organisms that encounter or colonize the leaf surface.[2] Leaf surface topography directly affects microhabitat[3] availability for colonizing microbes, microhabitat[3] quality[3] and acceptability for insects, and the efficacy of agricultural spray applications[4]. Prior[5] detailed[6] mechanistic[7] studies that examined particular[8] fungi-plant and pollinator-plant interactions have demonstrated[9] the importance of plant surface topography or roughness[10] in determining the outcome of[11] the interactions. Until now, however, it has not been possible to measure accurately the topography - i.e., the three-dimensional structure[12] - of such leaf surfaces or to record precise changes in patterns of leaf surface elevation[13] over time.

[contd]

COMMENT: What this abstract mostly needs is simplification, with concomitant abbreviation, following Style Guideline 1 (SG1). Only when another guideline is invoked is this marked. The passage also illustrates a new, modern tendency: to use "politically correct" terms, quite unnecessarily, either in a desperate attempt to engage the sympathies of the ecologically motivated or because this is how those who have gone into scientific research because they are ecologically motivated feel impelled to speak and write. The improved version on p 143 contains 68 words as against the 110 words above, yet is no less emotive.

1 **Landscapes** — unnecessary colorful, human-scale image.
2 What does this sentence introduce, other than a trendy phrase (**ecologically relevant**) to tack onto the human-scale image of a landscape? Delete.
3 Another unnecessary, *fashionable* word.
4 What does this word add? (SG2)
5 Malapropism for **previous** or **earlier**. **Prior** should be used sparingly, only when some action necessarily and immediately precedes another, and is usually correctly followed by **to**. (SG2)
6 Unnecessary adjective. Most studies are detailed.
7 Another fancy adjective. Presumably, the studies were of the mechanism of interaction, but they were not mechanistic.
8 Unnecessary adjective. The studies would not have been other than particular.
9 Verb beloved by the pompous which rarely means more than **shown**. When it does, use it by all means.
10 Topography *or* roughness? Choose the specific, not the abstract, noun here.
11 **Determining the outcome of = affecting**.
12 The number of times we encounter an explanatory phrase or synonym for **topography** makes one suspect that the authors don't have a very strong hold on its meaning.
13 Here's yet another one.

Improved botany abstract, part 1

MAPPING LEAF SURFACE CONTOURS

ABSTRACT The topography of a leaf surface affects its acceptability to colonizing microbes and insects and the efficacy of agricultural sprays. Earlier studies of interactions of fungi and pollinators with plants have shown how important plant roughness is to such interactions, but[1] it has not previously been possible to determine[2] the 3-dimensional structure of leaf surfaces or to record changes in it over time.

[contd]

1 Simplification of the preceding sentence allows **but** to be substituted for **however**.

2 **determine . . . the structure** is more accurate than **measure . . . the topography**. Can one measure a complex concept?

A pompous abstract in botany, continued

Using contact mode atomic force microscopy, we measured three-dimensional coordinates of upper leaf surfaces of *Vaccinium macrocarpon* (cranberry), a perennial plant, on leaves of two age classes[14]. We then produced topographic maps of these leaf surfaces, which revealed striking differences between age classes of leaves:[15] old leaves have much rougher measurements[16] than those of young leaves. Atomic force microscope measurements[17] were analyzed by lag (1)[18] autocorrelation estimates of leaf surface by age class. We suggest that the changes in topography[19] result from removal of epicuticular lipids and that the changes in leaf surface topography influence phylloplane ecology[20]. Visualizing and mapping[21] leaf surfaces permit detailed investigations into leaf surface-mediated phenomena[22], improving[23] our understanding of phylloplane interactions.

14 Do we need **classes**? Weren't they just of different ages?

15 The whole of the sentence up to here is redundant: the part after the colon says all there is to say.

16 Is it the measurements that were rougher? (SG2)

17 Unnecessary repetitiveness.

18 Abstracts should never contain a citation — they must stand alone (since they will appear divorced from the reference list in searchable services). This one is particularly obscure, with the citation number after the first word of a noun cluster.

19 What other changes have been discussed? Delete.

20 When one examines the meaning (SG2) of the fancy word **phylloplane**, one realizes it can only mean **leaf surface**. So the sentence boils down to **changes in the contours of the leaf surface affect the environment of the leaf surface**, which says nothing, or, to be more charitable, **changes in the contours of the leaf surface affect what happens there on the micro scale**, which is hardly amazing. Another sentence that can go.

21 Visualizing *and* mapping? Is there a difference? (SG2)

22 **leaf surface-mediated phenomena** is a noun cluster (SG3) that is just about bearable (only three elements), but would be equally succinctly expressed as **phenomena mediated by leaf surfaces**; then, one begins to wonder if **mediated** isn't too active a word, implying some sort of divine intention — surely **affected** is all one can say.

23 Dangling participle. (SG2)

Improved botany abstract, continued

Using[3] contact-mode atomic force microscopy, we[3] obtained 3-dimensional coordinates of the upper leaf surface of *Vaccinium macrocarpon* (cranberry), a perennial plant, for leaves of different ages. Old leaves were far rougher than young ones. We suggest that the changes result from removal of epicuticular lipids. Investigation of phenomena affected by leaf surfaces will be[4] facilitated by this kind of mapping.

3 At last, a correct usage in the original: **using** is not a dangling participle because the personal pronoun **we** is its subject.

4 More modest claim, putting the use of the new technique more clearly in the future.

Editing assignment 5
Part of a paper in radiation biology, (a) Title and abstract

EFFECT ON VIRUS-LIKE ELEMENTS OF EXPOSURE TO LOW- AND HIGH-LET RADIATION[1]

ABSTRACT

Changes in virus *expression[2] have been reported following *exposure to cellular stresses such as ultraviolet radiation or heat shock. The experiments reported here were designed[3] to examine the *expression[2] of endogenous virus-like 30S (VL30) elements of DNA after *exposure to gamma rays (300 cGy) or neutrons (50 cGy) at doses that are equally efficient in *cancer-production[4]. Gamma rays at 10 min post-exposure[5] induced[6] VL30 expression in gut and to a lesser extent in brain and liver. Neutrons repressed[7] VL30 RNA *accumulation[8] within 10 min following[9] *exposure in all tissues[10] (brain, gut, thymus, and spleen) except liver, in which VL30 RNA[11] was unaffected. These experiments suggest a differential transcriptional response following[9] *exposure to these two types of radiation.

NOTES ON EDITING ASSIGNMENT 5

The science here is also dreadful. I hope your students produce a better first draft than this. The paper does attempt to express things simply (SG1), but so tediously! Extreme tedium also results from a plethora of abstract nouns (SG3), here marked with asterisks.

1 Uninformative title (effect on what ?). Abbreviations not permitted in titles (what's LET ?). (SG2)

2 It's not the expression that changes, it's the rate or degree of expression. (SG2)

3 What's the purpose of **designed**? Either they (the experiments) or we (the authors) succeeded in examining this, or the paper is not worth publishing.

4 What's the purpose of the hyphen?

5 Ridiculous circumlocution for **after 10 min**: **post** is a warning word, to be used rarely, and only as a hyphenated preposition in a compound adjective.

6 **Induced** means **stimulated into production from a pre-existing or** [more usually] **zero baseline**. The context shows that **increased** is the right word. (SG2)

7 **Suppressed** (pushed down) or, if the decrease is only partial, **depressed,** would be better than **repressed** (pushed back).

8 Who said anything about accumulation? Always focus on what was actually measured (in this case, RNA concentration in tissue).

9 Warning word: **after** usually better, because it doesn't conjure up the image of people or things following something about like a dog.

10 Are these the only tissues there are? Should be **all tissues (we) examined**.

11 It was not the RNA that was affected, but the concentration of RNA. (SG2)

Improved version, Title and abstract

EXPOSURE OF [SPECIES] CELLS IN CULTURE TO GAMMA-RADIATION OR NEUTRONS INCREASES PRODUCTION OF VIRUS-LIKE DNA ELEMENTS SIMILAR TO THOSE INDUCED BY TUMOR PROMOTERS[1]

Key words: VL30, cancer, ionizing radiation, gut, liver

ABSTRACT

Endogenous production of virus-like DNA elements increases after cells are exposed to UV radiation or heat shock. We[2] report here that irradiation of [species] cells in culture with 300 cGy of gamma-rays also increases the production of virus-like DNA elements of density 30S but 50 cGy of neutrons suppresses it, although these amounts of ionizing radiation are equally efficacious at producing cancer. Percentage increases or decreases were different in cells from different tissues (brain, gut, liver, thymus, spleen).

NOTES ON IMPROVED VERSION OF EDITING ASSIGNMENT 5

The best way to begin is by combining the tables (p 154) into one. This gives insight into what the results actually were. Then work on the title and abstract.

1 It is no coincidence that the title has become longer and the paper shorter. Note this example of a title that is a statement.

2 Some journals forbid the use of the word **we** in abstracts, on the grounds that abstracts must stand alone. But since they always accompany the title and list of authors, I can't see the objection: the antecedents of the personal pronoun are unambiguously identified.

Note that the abbreviation VL30 has not been introduced in the improved title or abstract; it is not necessary in the abstract (because repetition of the term has been eliminated) and it would be unnecessarily distracting, especially as the reader would be left wondering whether VL30S or VL-30S wouldn't be better. If VL30 is a known abbreviation, fine, but unconventional abbreviations (i.e., those other than DNA, RNA) should be minimized in abstracts and totally avoided in titles.

Editing assignment 5, (b) Introduction and Methods

INTRODUCTION

[12]Virus-like 30S (VL30) elements[13] are genetic elements that are induced following[9] *exposure of cultured cells to tumor promoters, growth factors, and carcinogens and that are expressed at higher levels in tumor cells than in normal cells. The work reported here was designed[3] to determine whether ionizing radiation is capable of producing a similar response. We demonstrate[14] the induction of VL30 elements following[9] gamma-ray *exposure but not following[9] neutron *exposure, demonstrating[14] a differential[15] molecular response of the tissues[16] to the two types of ionizing radiation.

METHODS

Irradiations[17]. Gamma-ray irradiations were performed[18] using[19] a ^{60}Co source to deliver 300 cGy in 12 min. Neutron irradiations were performed[18] using[19] the Janus reactor to deliver a dose of[20] 50 cGy in 2 min.

RNA preparations[17]. Tissues were homogenized in a blender, extracted[21] with phenol, and precipitated[21] overnight in ethanol at -20°C. The RNA was then separated from DNA by precipitation[22] with 3 M sodium acetate, pH 6.0. Samples were treated with RNase-free DNase to remove contaminating DNA. *[NB: Details of the type of cell that was cultured and the culture conditions are omitted from this exercise.]*

12 It is conventional *not* to indent the first paragraph after a heading, but this is something a copy editor would take care of.

13 It is awkward to define an abbreviation in the middle of a set of stacked modifiers. Better: **Virus-like elements of DNA with a sedimentation rate** [I presume] **of 30 Svedberg units (VL30) . . .** (SG4)

14 Warning word: **show(ed)** almost always better. Here, the authors have become so enamored of it that they have their demonstration demonstrating, later in the sentence.

15 Warning word: does this simply mean **different**?

16 Are we talking about cells in culture or the tissues of a live animal? We still don't know. If the former, from what species are the cells derived? Presumably this was clarified in the part of the Methods section that has been omitted. However, all this information should be in the title.

17 It is not usual to put an abstract noun into the plural.

18 Warning word. Note how the use of this colorless verb in the passive voice allows the author to avoid saying what was irradiated.

Improved version, Introduction and Methods

INTRODUCTION

Virus-like elements of density 30S (VL30) are DNA elements produced in cultured cells that have been exposed to tumor promoters, growth factors, and carcinogens. They are expressed at higher concentration in tumor cells than in normal cells. We investigated whether ionizing radiation also induces VL30 elements and found that gamma-rays do but neutrons do not.[3]

METHODS

Irradiation. Gamma-rays (300 cGy in 12 min) were delivered by a ^{60}Co source. A Janus reactor was used to deliver neutrons (50 cGy in 2 min).

Preparation of RNA. Tissues were homogenized in a blender and extracted with phenol, and substances other than nucleic acids were precipitated overnight in ethanol at -20°C. The RNA was then separated from DNA by precipitation of the supernatant with 3 mol/L[4] sodium acetate, pH 6.0. The precipitate was treated with RNase-free DNase to remove contaminating DNA.[5]

3 It is unusual, though not forbidden, to announce detailed conclusions in the Introduction.

4 The designation M for **molar** is outdated; the above is the correct form.

5 I have made some assumptions in filling in just what was precipitated and what treated, which may be wrong, just as the reader may make the wrong assumptions if the information is not provided.

Notes from p 148, contd

19 Very common form of dangling participle. (SG2)

20 Non-parallel construction: why do we need **a dose of** here, but not in the first half of the sentence?

21 Lazy series of past participles. Were the tissues extracted and precipitated? That's what the sentence says. (SG2)

22 Of what? (SG2)

Editing assignment 5, (c) Results Part 1

RESULTS

The results of RNA dot-blot experiments designed[3] to quantitate (based on[23] microdensitometry) VL30 RNA levels in tissues are given[24] in Table 1. These results[25] show the induction of VL30 transcripts within the first 10 min following[9] *exposure to 300 cGy of gamma rays. Although induction was increased[26] in liver compared with[26] other tissues, the absolute levels of VL30 RNA in irradiated gut was[27] actually much higher; however, actual levels of induction (rather than the total amount) were not as great due to the high level of VL30 RNA already expressed in untreated gut.[28] In brain and liver, induction was greater at 60 minutes than after[29] 10 minutes following[9] irradiation. Thymus and spleen were also evaluated[30] for levels of VL30 RNA relative to controls, but due to the high levels of VL30 normally evident in these tissues, all *exposures were out of the linear range of the film (data not shown).[31]

23 Warning words, usually best turned into **on the basis of**, but here, **by** would do. The other fault is clumsy placement of the parenthesis, which interrupts the flow.

24 Warning word. Almost always, change **are given/shown in Table 1** to **Table 1 shows.**

25 Repetition of **results** should be unnecessary. (SG1)

26 Was it the induction that was increased? (SG2) Or the VL30 RNA concentration? Stick to what was observed. Or did the awkward phrase **increased compared with** mean, as usual, **greater than**?

27 Non-agreement of subject and verb is due to a failure to check the meaning and context of the word **was.** (SG2)

28 This extremely muddled sentence is the outcome of some very muddled thinking; the concept is simple and familiar enough. Replace with a statement of the observed facts.

29 Failure to realize that **following** means **after** has led to an absurd duplication of the word.

30 Warning word. In some European languages, **evaluate** simply means **measure** (i.e., determine the magnitude of), and this meaning has now corrupted scientific English. The authors did not determine the value of the thymus and spleen, but they measured something — not the thymus or spleen, but the VL30 RNA concentration. Increasingly, one longs to be told the units of this measurement — micrograms or picomoles per gram of wet tissue, perhaps? Far better than the relative values we are given.

31 The *exposures* were out of the linear range? (SG2)

Improved version, Results Part 1

RESULTS

Table 1 shows the results of RNA dot-blot experiments which quantified by microdensitometry the amounts of VL30 RNA in irradiated tissues. VL30 RNA increased in brain, liver, and gut 10 min after the gamma-irradiation. Its concentration had continued to increase in brain and liver after 60 min, but had declined in gut to the level observed before irradiation. Although the relative increase was greatest in liver at 60 min, the absolute amount of the increase was greatest in gut at 10 min, because this tissue normally contains so much VL30 RNA.[6] The amounts in unirradiated thymus and spleen are greater still, and the degree of induction could not be estimated because the densitometric film was totally black.[7]

6 Surely a reason for showing the absolute quantities rather than these ratios.

7 So why didn't the authors take smaller quantities of the tissue, or a weaker radioactive label?

Editing assignment 5, (d) Results Part 2

Similarly, Figure 1 and Table 2 show[32] the representative dot blot and microdensitometry results from experiments designed[3] to quantitate VL30 RNA levels in tissues following[9] *exposure to 50 cGy of neutrons.[33] These results demonstrate[34] repression[7] of VL30 RNA within[35] the first 10 min following[9] irradiation, which was observed in all tissues examined except liver, which showed no change in VL30 RNA level at the time points examined. For all other tissues, repression[7] of VL30 RNA was greater at 10 min than at 60 min following[9] neutron *exposure. Northern blots examining VL30 RNA levels in tissues 60 min following[9] irradiation similarly revealed an increase of VL30 RNA following[9] gamma-ray *exposure and a decrease in VL30 RNA following[9] neutron *exposure.

32 Readers of this book will long have realized that Tables 1 and 2 should be combined; further, is it likely that such a simple set of results, with so many of them close to zero, would benefit from being shown in a figure? (Step 11)

33 Unnecessary repetition, not only of what has gone before in the text but of the heading and footnotes to the table. (Steps 2, 11, 16)

34 Warning word: use **show**.

35 What do we know of **within**? Measurements were made only after 10 minutes.

Improved version, Results Part 2

The results with neutrons were different (Table 1): VL30 RNA production was completely suppressed 10 min after irradiation, and only partially recovered after 60 min – except in liver, which showed normal concentrations after 60 min.

Northern blot determination of VL30 RNA concentration confirmed the increase in VL30 RNA after gamma-ray exposure and decrease after exposure to neutrons.

Editing assignment 5, (e) tables as provided

Table 1. Relative VL30-RNA expression after gamma-ray exposure[a]

		Time after exposure (min)	
Tissue[b]	0	10	60
Brain	1.0[a]	2.9*[c]	4.9*
Liver	1.0	1.7*	8.7*
Gut	1.0	1.8*	1.2

[a] Expression of VL30 RNA was separately set at 1.0 for each type of tissue from controls; all other values are expressed relative to that. Standard deviations for all readings were less than or equal to 0.1.

[b] Thymus and spleen normally express high levels of VL30 RNA, and densitometric readings were impossible because all exposures were out of the linear range of the film.[31]

[c] An asterisk (*) indicates significantly different from untreated control at $p<0.001$.

Table 2. Relative VL30-RNA expression after neutron exposure[a]

		Time after exposure (min)	
Tissue	0	10	60
Brain	1.0	<0.1[b]	0.5*[c]
Liver	1.0	ND[d]	1.1
Gut	1.0	<0.1[b]	0.2*
Thymus	1.0	<0.1[b]	0.3*
Spleen	1.0	<0.1[b]	0.1*

[a] Expression of VL30 RNA was arbitrarily set at 1.0 for each type of tissue from controls; all other values are expressed relative to that. Standard deviations for all readings were less than or equal to 0.1.

[b] Value is less than 0.1, but absolute levels were below the sensitivity of the microdensitometer.

[c] An asterisk (*) indicates a value significantly different from untreated control at $p<0.001$.

[d] ND = not done.

Editing assignment 5
Improved version, tables combined

Table 1. VL30-RNA per wet weight of tissue after irradiation, relative to that in untreated tissue

Tissue	Time after exposure (min)	
	10	60
	After 300 cGy of gamma-radiation	
Brain	2.9[a,b]	4.9[b]
Liver	1.7[b]	8.7[b]
Gut[c]	1.8[b]	1.2
	After 50 cGy of neutrons	
Brain	<0.1	0.5[b]
Liver	ND[d]	1.1
Gut	<0.1	0.2[b]
Thymus	<0.1	0.3[b]
Spleen	<0.1	0.1[b]

a SD ≤ 0.1 for all readings **[*n* = ?]**.
b Significantly different from control at $P < 0.001$.
c Initial high concentration in thymus and spleen precluded measurement.
d ND = not done.

NOTES ON ORIGINAL TABLES 1 AND 2, OPPOSITE.

Table 1
The two tables should be combined, as above.
Footnote a is unnecessary given the column showing 1.0 at zero time. Both the column and the footnote should be deleted, as the title of the table says it all.
Footnote b is unnecessary, as the point has been made in the text.
Using one superscript (c) to refer to another (*) is astounding.

Table 2
Footnote b adds nothing.
Footnote d: one wonders why not. The lack of this datum means that the statement in the text is not supported.

Editing assignment 6
A complete short article

The synthetic short article that follows is not (apart from the table) faulty in logic or structure. Tell students to examine it as recommended for the revision of their own first draft, applying all four style guidelines. The superscript numbers should alert them to some faults, but there are others. The table is faulty in logical arrangement and in other ways. Provide the footnotes only afterward.

THE PROBLEM[1] OF LIPID PEROXIDES AS ARTEFACTS IN HUMAN AORTA LIPIDS
by W H O Dunnett and Slipshod M

Far Out University, No-man's Land

It has been suggested by some researchers[2] (1) that peroxides of lipids that are formed in the arterial wall conceivably[3] have a possible[3] role in causing[4] atherogenesis.This postulate has been widely accepted as a reasonable one, in view of the ready capability of breakdown of these compounds with the initiation of progressive chain reactions and the formation of a variety of[5] potentially toxic secondary[6] products.

In this connection[7], the destructive effects of lipid peroxides on serum lipoproteins (2) and on the SH group in proteins (3) have been noted[8]. Lipid peroxides appear in rat adipose and muscular tissues under conditions[9] of vitamin E deficiency and during the application[10] of a diet rich in polyunsaturated fats (4), and the formation of 'ceroid' in atherosclerotic arteries has been attributed to the autoxidation[11] and polymerization[11] of the unsaturated lipids they contain (5).

GENERAL NOTE: The author has applied "right justification" inconsistently (i.e., to some lines and paragraphs, not others). It is best to avoid a justified right margin altogether, because when the article is eventually printed the printer's software will be different and will make its own line breaks.

1 Are peroxides a **problem?** (Style Guideline 2)
2 Redundant. (SG1)
3 Too many cautious reservations. (SG1)
4 Omit (NB meaning of **atherogenesis**). (SG2)
5 **various.** (SG1)
6 Meaning? (SG2) Omit.
7 Weak connective. (SG1) Need a new paragraph here?
8 Avoid. (SG3)
9 Warning word.
10 Does one **apply** a diet? (SG2)
11 Abstract nouns. (SG3)

Editing assignment 6
An improved version

There follows one possible version of the faulty article, in which the four style guidelines have been applied. Other equally acceptable versions could be devised.

LIPID PEROXIDES IN EXTRACTS OF HUMAN AORTA
MAY BE ARTIFACTS

Doon Better
Far Out University, No-man's Land
FAX (+345) 123 246 8053, e-mail dbetter@ufarout.ac

It has been suggested (1) that lipid peroxides formed in the arterial wall are[1] atherogenic. The[2] suggestion[3] has been widely accepted as reasonable, because these compounds readily initiate chain reactions[4] to form toxic products. For example, lipid peroxides[5] denature[6] serum lipoproteins (2) and attack the SH group of proteins (3). When vitamin E-deficient rats are fed[7] a diet rich in polyunsaturated fats, lipid peroxides appear in their adipose and muscular tissue (4); similarly, it is thought, unsaturated lipids present in atherosclerotic arteries may autoxidize[8] and then polymerize[8] to form[8] "ceroid"(5).

1 Greater clarity (SG1), but no decrease of caution.
2 Eliminate **this** where possible. (SG1)
3 A **postulate** is something assumed without proof. (SG2)
4 A chain reaction is always **progressive**. (SG2)
5 No change in direction of ideas, therefore no need for new paragraph.
6 When the trapped verb in **destructive** is released, greater precision demands the more specific **denature**.
7 Much to be preferred (SG1) to the often-used **administered** unless it is by stomach tube.
8 Note the increase in simplicity and vigor when the trapped verbs are released. (SG3)

Following[12] the results of Lufton and Sowerby (1), the atherogenic role for lipid peroxides has been on firmer ground[13]. They[14] found an increasing[15] degree of lipid peroxidation in human aortic wall, positively correlated with the degree of atherosclerosis. It struck[16] us that the lipid extracting method[17] used by them, namely mixing with anhydrous sodium sulfate by means of a chop-knife and extraction of the mixture with chloroform, was liable[18] to effect[19] the artifactual formation of lipid peroxides from unsaturated lipids, and we undertook reinvestigation[20] of the lipid[21] peroxide[21] occurrence[21] problem employing extraction[20] at greatly reduced[22] temperatures in order to achieve[23] avoidance[20] of that possibility[20].

MATERIALS AND METHODS

Aortas were obtained at 6-12 h autopsy[24] at the University Hospital, Barchester. The classification of each as regards stage of atherosclerosis (0, I, II or III) was done[23] as described by Gee et al. (6), examining[25] the appearance of the intima, after which the adventitia (outer layer)[26] was rapidly but carefully[27] removed.

12 Avoid. (SG2)
13 Mixed metaphors (**role** on **firmer ground**?).
14 What is the antecedent? (SG2)
15 Continuous process? (SG2)
16 Too colloquial for formal writing.
17 Nonsensical construction: the lipids do not extract the method.
18 Only people are liable.
19 Colorless verb. (SG1)
20 Abstract noun. (SG3)
21 Modifiers stacked (SG4) around the foolish **problem**.
22 **much lower**. (SG1)
23 Colorless verb. (SG1)
24 A cluster that is both ugly and ambiguous.
25 Dangling participle. (SG2)
26 Explanatory phrase probably inappropriate for the audience.
27 Shouldn't all scientific work be careful?

Lufton and Sowerby (1) provided some evidence[9] for the atherogenic role of lipid peroxides. They showed that the peroxide content[10] in lipids extracted from the human aortic wall[11] correlated[12] with the degree of atherosclerosis. They[13] extracted the lipids, however, by mixing the tissue[14], which had been exposed to air[15], with anhydrous sodium sulfate and extracting the mixture with chloroform at room temperature[15]. These treatments may have caused the artifactual formation, by oxidation[16], of peroxides from unsaturated lipids during the extraction. We have therefore reopened the question of whether lipid peroxides[17] occur in aorta lipids, using anhydrous aerobic extraction at much lower temperatures in order to minimize oxidation.

MATERIALS AND METHODS

Aortic tissue[18]

Aortas were obtained at autopsy, within 6-12 h of death[19], at the University Hospital, Barchester. Aortas were classified[20] as being at stage 0, I, II, or III of atherosclerosis[21] according to the appearance of the intima (6), and the adventitia was rapidly removed.

9 The cliché **on firmer ground** replaced by neutral **evidence**.
10 Avoids abstract **peroxidation** (SG3) and speaks of **peroxides**.
11 Phrase has been expanded for greater precision.
12 **was correlated** would be equally acceptable.
13 Here, **They** can open a sentence because the antecedent is unambiguous. Moreover, the sentence construction is parallel to the preceding one.
14 **the tissue** added for clarification. (SG2)
15 **exposed to air** and **at room temperature** have been added to bring out why the method might have caused oxidation.
16 Specific and informative expansion of **formation**.
17 Breaking up the noun cluster (SG4) revealed that the **problem** is actually a question; releasing the hidden verb **occur** (SG3) further simplified the passage. Note that the usually colorless verb **occur** is here accurate and meaningful.
18 Subheadings inserted as signposts.
19 The shorthand jargon **6-12 h autopsy** has been expanded to be comprehensible and precise. (SG2)
20 Hidden verb **classify** released; colorless verb **done** eliminated. (SG3)
21 Phrase condensed by elimination of unnecessary words. (SG1)

The intima and media aorta preparations[28] were immediately put into a mixture of methanol and chloroform, approximately[29] 100 volumes, 2:1 (v/v). Storage[20] took place at -20°C. Following[30] the lapse of a given time (a period of one week was approximated[31]) the lower chloroform containing layer[32] was separated and analysis was performed[19] on it.

In some cases[33] the aortic material was further extracted with the same solvent at room temperature with stirring, and subsequently at the reflux temperature. The further yield of lipid was never more than half that initially extracted; the peroxide value of this extract was negligible, thus proving[34] that extraction at the lower temperature did extract peroxides effectively.

To determine[35] the oxidizability of lipid while it[36] is still contained in the tissue, a few aorta preparations were halved and one part exposed[37] to the atmosphere at room temperature (25°C) for 30 min before extraction, while[38] the other was extracted immediately after removal of the adventitia.

28 Noun cluster. (SG4)
29 **about**. (SG1)
30 **After**. (SG1)
31 **approximate** means **approach to a value**. (SG2)
32 The layer contains the chloroform, not vice versa.
33 Warning word: **cases**, ambiguous in medical work. When it means **patient**, say so; otherwise change to **instances** or some other unambiguous expression.
34 Dangling present participle (subject **these findings**). (SG2)
35 Dangling infinitive (subject **we**, not **aortas**). (SG2)
36 Careful! Is antecedent ambiguous? Here, slightly. (SG2)
37 Incorrectly omitted auxiliary (**one part were exposed**?).
38 Correct use of **while** — true simultaneity. (SG2)

Each preparation, comprising intima and media[22], was immediately submerged[23] in about 100 volumes of methanol-chloroform 2:1 (v/v)[24] and stored[25] at -20°C. After about 1 week the lower (lipid-containing)[26] layer was separated and its peroxide content was determined[27].

Some of the aortic material[28] was further extracted with the same solvent mixture[29] at room temperature with stirring, and subsequently under reflux[30]. The further yield of lipid never exceeded[31] 50% of that initially extracted; the lipid peroxide value of the second extract was negligible. These findings quashed the possibility that peroxides were not efficiently extracted at the lower temperature and permitted conclusions to be drawn from examination of the low-temperature extract exclusively[32].

To determine[33] how easily lipid is oxidized while it is still contained in the tissue, we divided a few aorta preparations into two, extracting[33] one part immediately after removal of the adventitia while exposing[33] the other to the atmosphere at room temperature (25°C)[34] for 30 min before extraction.

22 More information given; also, more precise in making clear that the preparation consists of one piece of tissue (therefore singular verb).

23 Descriptive, more precise verb replaces colorless one.

24 The word order now clarifies the relationships. (SG2)

25 Brevity resulting from release of hidden verb **store.** (SG3)

26 This explanatory parenthesis might not be necessary for experts in the field, but helps others without being offensive.

27 Abstract noun **analysis** replaced by precise phrase. (SG3)

28 One way to eliminate **case.**

29 Slight increase in precision. (SG2)

30 Slightly more concise. (SG1)

31 Precise, meaningful verb has replaced three colorless words.

32 Expansion for clarity (brevity is not always ideal). Note that some results are mentioned in this section. If necessary to justify later experiments or methodology, this is acceptable, indeed desirable.

33 This infinitive and these participles do not dangle.

34 **at 25°C** would be briefer, but would omit the implication that other investigators inadvertently allowed oxidation to take place. What **room temperature** is must always be specified, as it varies from place to place.

Aliquots[39] of the extracts were taken to dryness in tared glass shells for analysis[40] of total lipid content[40] and determination of peroxide contents were[41] conducted on similar aliquots[39].

Iodimetric lipid peroxide determination[42] is conveniently precise, and accurate for samples containing more than 50 µeq of peroxides (7), but as generally applied it is too insensitive for aortic extracts[43]. The microiodimetric procedure of Proudie and Slope (8) is approximately 2 orders of magnitude[44] more sensitive in comparison to earlier methods and enabled[45] results to be obtained on extracts from single aortas. The procedure was as described (8)[46].

RESULTS

The results are shown in Table 1. The number of values obtained for less diseased aortas is small, for the reasons given below. Examination of the table reveals that[47] the peroxide values obtained[48] are all very[49] much lower than those of Lufton and Sowerby (1), which ranged from 3, at their so-called[50] Stage I of atherosclerosis, to 17, at their Stage V, ueq/g lipid[51]. None[52] of our values exceeded 2 and most of them[53] were less than 1 µeq/g.

39 An aliquot is an exact fraction of the whole; the correct word here is **sample**, though in fact neither is needed (see improved text).

40 Not **analysis**, for no subfractionation is implied; **determination** or **assay** would be correct. (SG2)

41 Incorrect number of verb (contamination by **contents**). (SG2)

42 Noun cluster; not too bad, but easily and neatly avoided. (SG4)

43 Not clear what's special about aortic extracts (the author meant it was too insensitive for extracts of single aortas). (SG2)

44 Expression to avoid. (SG1)

45 Best to **enable** people, **allow** things to happen. (SG2)

46 State the principle of a quoted method.

47 All words in the sentence preceding this are redundant. (SG1)

48 Redundant word. (SG1)

49 Overworked word, usually avoid. (SG1)

50 This is derogatory. Correct term is **what they called Stage I**.

51 Far too many commas, and they do not save the clumsy word order.

52 Table 1 shows this not to be true.

53 **of them** redundant. (SG1)

Total lipid concentration of each extract was determined by drying[35] a portion in a weighed glass shell; the peroxide concentration of another portion was measured[36] by micro-iodimetry[37].

Determination of lipid peroxides by micro-iodimetry[38]

The iodimetric method for determination of lipid peroxides is convenient, precise, and accurate for samples containing more than 50 µeq of peroxides (7), but as generally applied it is too insensitive for use with extracts of single aortas[39], which usually contain (especially if they are fairly free from atherosclerotic lesions) a small total amount of lipid which itself contains little peroxide. The micro-iodimetric method of Proudie and Slope (8) is about 100 times[40] as sensitive as earlier methods and was successfully applied. It employs electrometric titration of iodine liberated from iodide by the peroxides in acid solution.

RESULTS

The results are shown in Tables 1 and 2.[41] The number of values obtained for less diseased aortas is small, because the latter could be obtained only from autopsy of infants or children.

35 Dangling participle of the least objectionable kind.

36 The hidden verb to be released was **determine**, but this has been used earlier in the sentence, and **measured** is substituted to avoid repetition.

37 More information, which provides a link with next section.

38 Subheading helps readers' scanning.

39 Inclusion of **single aortas** clarifies the thought.

40 Straightforward factor instead of **order of magnitude**. (SG1)

41 The data have been separated into two tables because there were two points to be made. See notes under the tables.

Table 1. Peroxide values

Tissue exposed to air[a]			Tissue not exposed before extraction		
Stage of athero-sclerosis	Tissue no.[b]	Peroxide content	Stage of athero-sclerosis	Tissue no.[b]	Peroxide content
		μeq/g			μeq/g
III	21	2.62	III	1-13,20,21,22	1.26±0.91[c]
II	22	1.64	II	14-17	0.84±0.30
I	23	1.81	I	18,23	1.18±0.57[c]
			0	19	0.41

Table 1 also shows that[54] when tissue was exposed to the air at 25°C the peroxide value was increased 2-3 times (tissues 21-23), suggesting[25] very[49] strongly that in this situation[55] lipid peroxides are formed through an artifactual conversion[56] of the lipids before they are extracted.

The design of the table is confused. When one tries to expand the title to be more informative it becomes obvious that two tables are needed, each to show a different point.

a Since one reads from left to right, it is normally better to show controls on the left. Similarly, as one reads from top to bottom, it is usually best to show controls, or less affected samples, at the top.

b The numbers designating individual samples of aortic tissue are of no assistance to readers unless they can compare values for any one tissue in different places in the table. Here, individual values have, in the right-hand half of the table, been buried in means.

c There is no statement as to whether the variation is SD or SEM. This must always be stated, together with the number of observations. When there are only two values, it is foolish to report them as x±y.

54 All words before this are redundant. (SG1)
55 Warning word: clarify meaning or omit. (SG1,2)
56 Circumlocution. (SG1)

NOTE ON THE IMPROVED TABLES (OPPOSITE)
The titles and layout expose experimental design completely to a reader who has not yet looked at the text. Important information is given succinctly in the footnote to Table 2.

Table 1. Peroxide values of lipids from aortas at different stages of atherosclerosis

Stage of atherosclerosis	No. of aortas	Peroxide content $\mu eq/g$
0	1	0.41
I	2	0.61, 1.75
II	4	0.84±0.30[a]
III	16	1.26±0.91[a]

[a] SD.

The peroxide values in Table 1 are all much lower than those of Lufton and Sowerby (1), which ranged from 3 (their Stage I) to 17 (their Stage V) μeq/g lipid. Only one of our values was higher than 2 and most were below 1 μeq/g. There was no correlation between peroxide content and stage of atherosclerosis[42].

Table 2. Effect on lipid peroxide concentrations of exposing tissue to air before extraction of the lipids

Aorta No.	Stage of athero-sclerosis	Peroxide content	
		Extracted at once	Exposed 30 min[a]
		$\mu eq/g$	
1	I	0.61	1.81
2	II	0.81	1.64
3	III	0.80	2.62

[a] At 25°C.

Exposure of the tissue to air at room temperature increased the peroxide value 2-3 times (Table 2), which strongly suggests[43] that lipid peroxides are easily formed artifactually[44] before the lipids can be extracted.

42 An extra, important conclusion has been added.

43 Replaces dangling participle (the commonest way of doing so).

44 Replaces circumlocution. (SG1)

Since some exposure cannot be avoided during manipulation[57] at autopsy and during removal of adventitia, all values in Table 1 are likely[58] too high. Based on[59] this result, and coupled[60] with the fact that even these sensitive methods do not have the ability[61] to yield an accurate result on the small amount of lipid obtainable from Stage 0 or Stage I aortas, the project of comparing peroxide contents of aortas with varying[62] degrees of atherosclerosis has been abandoned at the present time[63].

DISCUSSION

If, as seems likely from these results, the high values of Lufton and Sowerby (1) for the aorta lipid peroxide content[64] was due to artifactual formation of peroxides [during tissue preparation and possibly during the extraction of the lipids][65], the question arises[66], why did these authors obtain a correlation of peroxide content with atherosclerosis? [At the time of their investigation (1951) it was not known that][65] the arterial lipid unsaturation index[64] increases with increasing degree of atherosclerosis. This increasing unsaturation, most striking in the cholesteryl esters, was shown by Gee et al. (6) [with the aid of gas-liquid chromatography][65]. Furthermore, the proportion of cholesteryl esters relative to the other lipids rises also, [as has long been recognized][65] (9,10). The atherosclerotic aorta therefore bears an increasingly oxidizable load, and this would be sufficiently explanatory[67] of the results obtained.

57 Redundant polysyllable. (SG1)
58 Better kept as adjective (as in **likely to be too high**); the correct adverb is **probably**.
59 Dangling past participle; replace by **On the basis of** if necessary. (SG2)
60 Again; but if **On the basis of** is substituted, becomes all right because **this result** is truly the subject of **coupled**.
61 **cannot**. (SG1)
62 **Varying** happens over time: correctly, **various**. (SG2)
63 Redundant. (SG1)
64 Noun cluster. (SG4)
65 There are few stylistic or grammatical errors in the Discussion, except for the roundabout phraseology and inclusion of irrelevances, marked by [ ].
66 Redundant. (SG1)
67 Hidden verb. (SG3)

Since some exposure is inevitable during autopsy and removal of adventitia, all values in Table 1 are likely to be[45] too high. For this reason, and because even these sensitive methods are incapable[46] of giving an accurate result on the small amounts of lipid that can be extracted[47] from Stage 0 or I aortas, the project of comparing peroxide contents of aortas with different[48] degrees of atherosclerosis has been abandoned.

DISCUSSION[49]

If the peroxides measured in lipid extracts from the arterial wall are artifacts, how can we explain Lufton and Sowerby's findings (1) that the peroxide content is correlated with degree of atherosclerosis? It has recently been discovered (6) that arterial lipids become progressively more unsaturated with increasing degree of atherosclerosis. Among the lipid classes, cholesteryl esters show the most striking increase in unsaturation, and the proportion of cholesteryl esters relative to the other lipids also rises (9,10). The more atherosclerotic the aorta, therefore, the more susceptible will its lipids be to oxidation during dissection; this effectively explains the observed correlation.

45 Note insertion.
46 Replaces roundabout phrase. (SG1)
47 More precise than **obtainable**.
48 Correction of **varying**.
49 The argument is tighter because the author has eliminated irrelevant information and connects the ideas more closely.

We should like[68] to point out, however, that lipid peroxidation may still be involved[69] somehow in the inception of atherosclerosis or in forwarding its progress. Although only small amounts of lipid peroxides were found, they[70] might be genuine components of the tissue. The possibility also exists that more were present in the living patient, but decomposed between the time of death and the autopsy. Their very decomposition in the living tissue might have been the cause of atherosclerotic changes[71]. But it is not our opinion that measurement of peroxides in necropsy material can elucidate this problem[72].

REFERENCES

1 L.[a] Lufton and Sowerby J, 1982.[b] *Acta.*[c] *pathol et.*[c] *microbiology*[d] 30:492[e].

68 Absurd conditional: the authors *are* pointing it out. (SG2)
69 Colorless verb. (SG2)
70 Check the antecedent (amounts? peroxides?). (SG2)
71 Circumlocution. (SG1)
72 Avoid (allows the authors to be vague; what, of the various speculative statements in preceding sentences, do they think is **the problem**?).

If you wish, you can use the editing assignments to provide instruction on the correct listing of references, in the style required by the target journal. In the listing on p 170, the references are clearly inconsistent with one another, and the authors have made no attempt to follow any Instructions to Authors. The references in the improved text opposite are in the "Vancouver style", now widely accepted by biomedical journals, and the footnotes keyed to the faulty reference list below relate to deviations from this style.

Make the point also that the list of references must be submitted to the journal double-spaced, exactly like the bulk of the text. The spacing on p 170 is variable.

a Initial should not precede surname.
b Almost all journals nowadays require the title of the article cited to be given in the reference; it is much more informative to the reader, and also provides a check that the citation numbering has not gone astray.
c The Latin words **Acta** and **et** are complete, not abbreviations, and therefore never take a period. The **et** is in any case redundant; no standard abbreviation of journal titles includes **and** or **&**.
d As the journal title is in Latin, this is evidently incorrect.
e Inclusive pagination almost always required.

Our results do not exclude the possibility that lipid peroxides play a role in atherogenesis or in the development of atherosclerosis. The small amounts found may not be entirely artifactual. Furthermore, lipid peroxides present in vivo may decompose between death and autopsy. More importantly, they may have formed earlier in the patient's life and subsequently decomposed, with the undesirable consequences mentioned in the Introduction. We do not believe, however, that their possible role in atherosclerosis has been or can be established by examination of the lipids after death.

REFERENCES

1 Lufton L, Sowerby J. Role of lipid peroxides in atherogenesis. *Acta Pathol Microbiol* 1982;30:492-5.

2 Robarts, M.,[f] O. Omnium and Dumbello, G. 1991. *J Perfect Res* 2:631.

3 Neroni, M.V. 1983. *Bull. Soc. ital. sper. Biol.*[g] 92:16

4 De Courcy, E. <u>Extrait de Bull. Soc. Chem.</u>[h] <u>France</u>[i] 116, 214-324, 1993.

5 Hartletop, M. O. J. Gerontol.[j] 63, 321 (1988).

6 Gee, N.B., Whiz, A.C. and W. A Goof <u>Scalpel</u> i:137[k], 1990.

7 Eaxfs, J. and Buffle, R. (ed.) *Organic peroxide analysis* i: 376, 1886.[l]

8 Proudice, M. B. & O. Slope. (1994). J. A. M. A.[m] 3:423.

9 Dunstable, E. et al.[n] *Biochem*[h] *Biophys Acta* 74:111, 1987.

10 Doppelganger[o] D, D Winterreise. *Z Phys*[p] *Chem* 216(1955)984.

f Note how much scope there is for confusion, should the comma be omitted, if the initials sometimes come after the surname, sometimes before.

g Arbitrary decisions on the capitalization of words in the journal title.

h Misspelled foreign word that is similar to its equivalent in English.

i Sudden switch to underscoring instead of italicization.

j Is this MOJ Hartletop in *Gerontologia*, or MO Hartletop in *J Gerontol*?

k Omission of last digit of page number, see opposite (this would not be confusing if inclusive pagination were given).

l A multitude of faults, including garbled spelling of first author, treatment of a book as a journal, grossly wrong date, omission of publisher's name and place of publication.

m Although the *Journal of the American Medical Association*, and a few other medical journals, permit the use of *JAMA* in references, it is forbidden in others, and it is never incorrect to use the proper abbreviation, *J Am Med Assoc.*

n Practice on the use of **et al.** differs in different journals, but all journals require at least three authors' names before **et al.** is used; some require six, and some bar the practice altogether.

o Foreigners like to receive all their diacritical marks, if the author's word processor can cope.

p Should be **Physiol**; this abbreviation stands for Physische (Physical), so the mistake could make retrieval of this article difficult or impossible.

2 Robarts M, Omnium DO, Dumbello G. Chain reactions initiated by peroxides in serum lipoproteins. *J Perfect Res* 1991;2:631-4.

3 Neroni MV. Denaturation of proteins by the action of lipid peroxides. *Bull Soc Ital Sper Biol* 1983;92:16-22.

4 De Courcy E. Generation of peroxides in rat adipose and muscular tissue fat by administration of polyunsaturated fats. *Bull Soc Chim Fr* 1993;116:214-30.

5 Hartletop MO. Ceroid in atherosclerotic arteries. *J Gerontol* 1988;63:321-6.

6 Gee NB, Whiz AC, Goof WA. Grading of atherosclerosis. *Scalpel* 1990;1:1378-9.

7 Eames J. Iodimetric method to determine lipoperoxides. *In* Buffle R (ed). Organic peroxide analysis, vol 1. Barchester: Cathedral Publishers Ltd; 1986. pp 376-8.

8 Proudie MB, Slope O. A micro method for lipid peroxides. *J Am Med Assoc* 1994;3:423-38.

9 Dunstable E, Gresham F, Thorne M. Composition of cholesteryl esters of the arterial wall in atherosclerosis. *Biochim Biophys Acta* 1987;74:111-21.

10 Doppelgänger D, Winterreise D. Arteriosklerosefetten. *Hoppe-Seyler's Zeitschr Physiol Chem* 1995;216:994-1015.

Note consistency of style, inclusion of titles, inclusive pagination of papers cited. Note too that the date of ref. 10 has been corrected (typing 5 for 9 is easily missed): the author often forgets to check the final version of the reference list, footnotes, tables, and legends.

Appendix 3

Warning words and phrases

Often, these are expressions to be avoided, *though not always and not at all costs.* The table gives some *suggested* substitutions. Most usages of these words or phrases are violations of Style Guideline 1, 2, or 3 — or all three of them! **The cardinal rule is: if you see this word or phrase in your draft, think hard about whether you are justified in using it.**

Avoid	*Usually* prefer
anticipate	expect
approximate	[use only to mean **approach asymptotically**]
approximately	about
as already stated	[omit]
as can be seen from Fig. 1, growth is more rapid	growth is faster (Fig. 1) *or* Fig. 1 shows that growth is faster
as far as our own observations are concerned, they show	our observations show
as of now	now, from now on
as shown in Fig. 11	Fig. 11 shows
at some future time	later
at the present moment, at the present moment in time, at this time	now
authors (the)	I/we
based on	on the basis of
case; not the case	patient; not so
color (red in color, etc.)	red, etc.
colorless verbs	replace by real, active verb
commence	begin
comparable	similar
conducted inoculation experiments	inoculated
considerable amount of	much
considerable number/proportion of	many/most
decreased number of	fewer
decreased relative to	less/fewer/lower than
demonstrate	show
due to	because
due to the fact that	because

Avoid	***Usually* prefer**
during the time that	while
elevated	raised, higher, more
employ	use
encounter	meet
encountered frequently	common, frequent
exhibit	show
fewer in number	fewer
following	after
for the reason that	because, since
greater/higher number of	more
if conditions are such that	if
in a considerable number of cases	often
in all cases	always
[adjective] in character	[adjective]
in comparison to	than
increased relative to	more/higher than
in excess of	more than, above
in spite of the	despite
in the present communication	here, in this paper
in excess of	more than
in terms of	[find the exact relationship]
in view of the fact that	since, because
it has long been known that	[omit]
it is of interest to note that	[omit]
it is often the case that	often
it is possible (probable) that	possibly, probably
it may however be noted that	however, but
it seems to the present writer	I think, we believe
it will be seen upon examination of Table 5 that	Table 5 shows that
it would thus appear that	apparently
large number(s) of	many
large proportion of	much, most
less	fewer [if qualifying a number, not an amount]
lesser extent, lesser degree	less
level	concentration, content, amount, degree
majority of	most

Avoid	***Usually* prefer**
moment in time	time
multiple	several, many, different
number of	several, some, many
of large size	large
of such strength that	so strong that
order of magnitude greater than	10 times as [etc.]
owing to the fact that	because
parameter	index, criterion, factor, measure, characteristic, value, variable
prior to	before
problem	[usually not a **problem**, but if there is one, define what it is]
proportion of	some
quite	[omit]
rather	[omit]
relatively	[avoid, unless you are calculating a real quotient]
respectively	[use only in this correct sense: **the values for A, B, and C were x, y, and z respectively**]
reveal	show
shape (square in shape, etc.)	square
significant	[avoid, unless referring to statistical significance (with defined *P* value), by using **much**, **appreciable**, **substantial**]
sophisticated	advanced, complex, new, expensive
sufficient number of	enough
subsequent to	after
terminate	end
there is, there are	[often unnecessary; always examine for possible omission of **there**]
they are both alike	they are alike
upon	on
using	[test whether it is a dangling participle; if so, try **by**, **with**, **by means of**]
utilize, utilization	use
very	[omit]
while, whilst	although
with regard to	in, to

Appendix 4

Points of grammar and punctuation that your students may have forgotten

In Chapter 7 reference is made to some grammatical terms that your students may have forgotten. This appendix does not provide a complete glossary of grammatical terms, but focuses on the relationships between the kinds of words you tell students to look out for under Style Guideline 2 ("Check the meaning and relationships of every word").

The remarks on pp 179–184 on punctuation focus on those a competent copy editor can *not* correct from sight of the manuscript alone, and which authors therefore need to understand as they revise the first draft.

Verbs

The most important relationship, and one that most often goes awry, is that between subject and verb. Verbs are best characterized as words of action or being. Their *indicative* forms indicate past, present, or future actions or states and must have a subject, either stated or implied. They can be in the active or passive *voice*; if the latter, the subject may be unstated, and in poor scientific writing it is often wrongly implied by the context (see table, p 177).

Past, present, and *future* are the names of the commonest *tenses* used in scientific communication. Findings or actions are usually reported in the past tense ("we **found**, the rats **swam**, the cages **were** disinfected, the yield of protein **was** . . ."). A property that is widely accepted usually takes the present tense ("the hardness of diamonds **is** due to, Jones et al. have shown that the stability of this compound **rests** on"). Also useful in scientific writing is the *pluperfect* or *past perfect* ("After the mixture **had been** centrifuged, cells were harvested from the precipitate").

Other forms of the verb are the *infinitive* (**to do**, **to be**) and *participles,* either past participles (**centrifuged**) or present participles (**centrifuging**).

Problem 1 with verbs

In English, verb *endings* don't change much, except for the distinction between singular and plural — and this is where things often go sour. If writers get into the habit of asking of every verb "What, actually, is the subject of this verb?", they

- will be following Style Guideline 2,
- will correct mistakes like "Lightening of the colors of the leaves are brought about by . . .", and

- will discover ambiguities and obscurities in sentences they have written in the heat of the first draft and will rearrange the words accordingly.

Problem 2 with verbs

The second major problem is the handling of the passive voice, which is used so much — some say, too much — in scientific writing. The passive voice requires use of an auxiliary verb, part of the verb **to be**, most usually in scientific writing **was** or **were**, depending on whether the subject is singular or plural. The commonest fault here is to list several actions, omitting the auxiliary verb after the first one even when the subjects change from singular to plural or vice versa.

Thus, while it is correct for James Bond to say his **"martini should be shaken, not stirred"** — because the subject (**martini**) of the two verbs **shake** and **stir** is the same — and while it is equally correct to say **"the cells were spun down, separated from the supernate, and resuspended in a different medium"**, again because the subject (plural this time) remains the same throughout, it is *not* grammatically correct to say that "the cells were spun down, the supernate decanted, and three portions of Eagle's medium brought to pH 6.2 added". The subjects have changed in number (**cells**, **supernate**, **portions** — plural, singular, plural). Furthermore, omission of the auxiliary verb is the cause of further confusion when an adjectival (qualifying) participle (**brought**) is thrown in, for the reader's first instinct is to see it as one of the passive verbs being listed. The best rendering of the passage would be **"The cells were spun down, the supernate was discarded, and three portions of Eagle's medium at pH 6.2 were added."**

Problem 3 with verbs

Another common fault is the *dangling* (or, more logically, *"unattached"*) *verb form*, where a sentence begins with a participle (present or past) or an infinitive that has an implied subject, but the sentence continues with a subject that is not the same. The table opposite gives examples, and here is another:

> After centrifuging the suspension, the cells were harvested from the particulate matter.

Here, the cells were certainly not doing the centrifuging, i.e. are not the subject of the preceding participle. So the writer must ask "What is the subject?" of *every* verb form, including — or especially — participles and infinitives, in order not to raise an unintended smile on the reader's face.

The fault is corrected in one of three ways:

- change the main verb into the active voice, naming the true subject, thus

"After centrifuging the medium, **we** harvested the cells [from the particulate matter] as follows . . .", or

- use the pluperfect instead of the participle, thus
  "After the medium **had been** centrifuged, cells were harvested . . .", i.e. naming the correct subject for each verb, or
- substitute an abstract noun for the verb form, thus
  "After **centrifugation**, the cells were harvested . . .",
  but Style Guideline 3 suggests you limit your use of abstract nouns that mask verbs, so use this solution only if it avoids clumsiness.

EXAMPLES AND NAMES OF VERB FORMS AND DANGLING VERB FORMS

(Dangling constructions, printed in italics, are not acceptable.)

Example	Name	Subject of verb
He claims	Present tense, active voice	He
It is claimed	Present tense, passive voice	It
Claiming to have the necessary evidence, this author states	Present participle	This author
Having claimed priority, Jones stated	Past participle	Jones
Having claimed priority, his grant application was turned down	Dangling past participle, **having**	*Not* his grant application
Claiming priority can be dangerous	Gerund (a verb form that is a noun)	Claiming priority (the verb in this sentence is **can**)
To claim priority, Tim needs dated evidence	Infinitive of purpose	Tim (the verbs are **claim** and **needs**)
To claim priority, dated evidence is necessary	**To claim** is a dangling infinitive	*Not* evidence; the subject of **To claim** is not stated
He was claiming They were claiming	Auxiliary verb (**was, were**) plus present participle	

Clauses and phrases

A clause is a string of words that hangs together to make sense as a whole. It must contain a verb in indicative form (past, present, or future tense, active or passive voice). A phrase is *also* a string of words that makes sense as a whole, but it does not contain a verb in indicative form. A phrase may never therefore constitute a sentence, which by definition must contain such a verb. A phrase may, and often does, contain a participle:

> She learned a lot, [clause, containing the verb **learned** in the past tense] attending those lectures diligently and taking copious notes [phrase, containing two participles].

and

> He went to all the lectures [clause] to gain as much knowledge as possible [phrase, containing a verb in the infinitive].

The skill lies in ensuring that the verb in the infinitive form is properly attached. In these examples, the subject of **attending** and **taking** is **she**, which is indeed the subject of the main clause; and the subject of **to gain** is **he**, ditto.

Modifiers and stacked modifiers

Modifiers are words that pin down a noun by specifying its particular characteristics. Modifiers that are *adjectives*, for example

> a pretty girl, tall girls, a handsome man, short men, Gram-negative bacteria, platitudinous introductions

give no particular trouble, even if there is a string of them before and even after the noun:

> **tedious, long-drawn-out, wordy** discussions **characteristic** of this author . . . .

Participles are often used as adjectives, again without noticeable problems:

> a shocking announcement; stunning blows; dragging, thumping, banging noises; a low-pitched sound.

However, English also allows *nouns* to function as adjectival modifiers without changing their form:

> heart valve, plant species, laboratory technician, farm hand.

This also raises no difficulties until the number of such modifiers becomes excessive (roughly speaking, more than two). Why? Because each word in a string of modifiers could, in principle, modify any of the others in the string, and this gives rise to ambiguity or, sometimes, laughter. Take our example in Chapter 7:

> heavy beef heart mitochondria protein

where the initial adjective is followed by four nouns. There is not much ambiguity about **beef heart mitochondria** — these are mitochondria from the hearts of oxen (or cows?). Purists might like to insert a hyphen, thus **beef-heart mitochondria**, so that **mitochondria** would then have only one modifier, but this is regarded as rather fussy punctuation nowadays. The string **beef heart mitochondria protein** begins to be ambiguous, or at least causes the reader to hesitate slightly over whether the protein is heart protein or mitochondrial protein, then to come down on the side of the latter. Note that adding a single letter, the *l* of mitochondrial, and insertion of that fussy hyphen to give **beef-heart mitochondrial protein** would make the meaning absolutely unequivocal.

What have we done? Reduced the number of modifiers from three to two, and turned one of them from a noun into an adjective. *Now,* the adjective **heavy** would clearly apply to **protein** and not to beef, heart, or mitochondria. Even so, the phrase would still be better expressed as **(the) heavy [chain of the] protein extracted from** (or **derived from/originating in/characteristic of**) **the mitochondria from ox hearts**. Notice that splitting up the stack of modifiers forces the writer to clarify exactly the relationships between the words, and that cannot be done unless and until the relationships are fully understood. So it's the author's job, not the reader's.

Punctuation

Period

Few have difficulty with this punctuation mark ("full stop" in British usage): it separates complete sentences, i.e. statements each containing at least one indicative verb (past, present, future tense, in active or passive voice). However, one common fault arises when a period concludes a complete sentence within parentheses. When this happens, the period must be placed before the concluding parenthesis, not after it.

Thus, in the previous paragraph, if the last sentence had been added as an afterthought and was therefore in parentheses (or brackets, as they are often called for simplicity, though not correctly), the punctuation should not be

> . . . complete sentence within parentheses. (When this happens, the period must be placed before the concluding parenthesis, not after it).

but

> **. . . complete sentence within parentheses. (When this happens, the period must be placed before the concluding parenthesis, not after it.)**

Parentheses

Over-use of parentheses gives the impression that the writer is either hesitant or too frequently given to second thoughts that are then thrown in pell-mell instead of being integrated into a rearranged paragraph. A writer who does use them must make sure the closing parenthesis is (a) present and (b) in the correct place.

The test for (b) is easy. Imagine the whole parenthesis deleted; if what is left is grammatically correct and makes sense, the parentheses are in the right place. Apply this test to the example of where the period should go in the previous section, and the correct placement is immediately apparent: if they had been wrongly placed, the passage after deletion of the parenthesis would have two periods, one after the other.

Commas

Coping well with commas is much more difficult. One cardinal rule is: never separate a subject from its verb by an odd number of commas. A *pair* of commas may separate off a clarifying phrase from the rest of the sentence, and separate subject from verb:

> **Jack and Jill, who were both wearing stout shoes, set off for a walk**

is correct, but

> The Agriculture Commissioners who were assigned to this task, were adamant in their demands

is wrongly punctuated. The writer has put in the interfering comma because of feeling mentally short of breath after the *defining clause* **who were assigned to this task**, but there should be no interruption between the subject, the **Agriculture Commissioners**, and its verb, **were**. If the sentence had been slightly different in meaning, a *pair* of commas would have been appropriate:

> **The Agriculture Commissioners, who were assigned to this task against their will, were adamant in their demands.**

Notice the change in meaning: now, the sentence indicates that all the Commissioners had been assigned to the task, whereas in the previous version only some of them had been.

This leads us straight into clarifying the distinction between *defining* and *non-defining* clauses/phrases. The first sentence about the Commissioners defined a group among them as those who had been given a task. The second sentence, with the commas, says something about all of them. Non-defining (or "descriptive") clauses like "who were wearing stout shoes" in the first example are preceded by a comma and end with one; the sentence would make sense if the whole clause flanked by the commas was omitted. Defining clauses are *not* preceded by a comma and must not end with one; omission of the defining clause would make nonsense of the sentence.

The same rules apply to phrases: the first sentence might just as well have been

Jack and Jill, both wearing stout shoes, set off for a walk

in which **both wearing stout shoes** is a descriptive phrase, not a clause; it is flanked by a pair of commas. By contrast, our first sentence about the Commissioners might have read

The Agriculture Commissioners assigned to this task were adamant in their demands

in which the phrase **assigned to this task**, as it is defining, must have no commas.

Attention to the unambiguous use of commas is far from being a trivial pursuit: the meaning of a passage can be profoundly altered by incorrect usage. When your students have grasped the difference between defining and descriptive clauses, together with the crucial role of commas in distinguishing them, they should readily perceive that the comma in the following should be deleted:

Under these circumstances lysosomal enzymes cannot acquire the mannose-6-phosphate recognition marker, which allows them to be properly routed to the lysosome

and should be ready to check for similar wrong usage in their own drafts. Even better would be to make **which** into **that** once the comma has been banished: **that** marks a defining clause quite securely.

Now let us revert to the commonest fault of all: insertion of an unnecessary and distracting comma, (like this one!) for no reason other than that the writer feels the reader needs a breather. Another reason (but no excuse) may be technical: the comma is a sort of tombstone, the sole remains, of a clause or phrase that has been deleted from the word-processor but without deletion of the introductory comma. This must be the explanation for the presence of a comma in this opening sentence in a 1997 letter to *The Lancet*, which is otherwise breathtakingly inexplicable:

Schizophrenia, is a complex genetic disorder and may involve multiple genes of small effect.

However, the pomposity of the rest of the sentence makes one wonder whether the explanation is too charitable; the combination of faults indicating muddled thinking certainly prevented one reader at least from pursuing the argument any further.

Colons and semicolons

A colon always points forward: the second part of the sentence (a complete clause, like the first part) either exemplifies or amplifies the clause before the colon. The colon is also used before a list, as in the following hierarchy of punctuation marks:

comma, semicolon, colon, dash, period.

Here again it points forward. Notice how I have used the colon in both the sentences so far in this section.

A semicolon does not point forward, although it too separates two complete statements, complete with verbs; use of a semicolon instead of a period indicates some degree of dependency, either of reinforcement or disagreement, between the two statements.

Semicolons and colons are *not* interchangeable. The former do *not* point forward and must not be used to indicate that a list follows. Words of affirmation or contrast like **furthermore**, **moreover**, **indeed**, **nonetheless**, or **however** are often added after a semicolon to reinforce the interdependence of the two statements, whereas such words would destroy the force of a colon if they were placed after it: the second statement after a colon must follow logically and without interruption directly from the first. If it does not, the colon has been misused.

One further use of the semicolon is to separate items in a list of actions or statements which themselves contain complex grammatical constructions, perhaps with pairs of commas within them, for example:

> The seedlings are first planted, 6 cm deep and 2 cm apart, in a strongly alkaline compost; bamboo rods impregnated with paraquat, the latter previously incubated at 25°C for 24 h in the dark, are placed alongside in parallel rows 10 cm apart; the whole is subjected to alternating light and dark cycles, 4 h each, for 8 days; and the leaf third from the soil surface is measured along its whole length.

Some purists might object to the length of this sentence, but I think you will agree that if properly punctuated it is perfectly clear, easily followed, and therefore eminently acceptable.

Hyphens

A modern tendency is for writers to omit hyphens from compound adjectives where formerly they were considered essential. As a result, some writers add them where they do not belong. A special favorite with these writers is "short-term".

The rule is really quite simple. When two words, of whatever kind, are joined to form what is known as a *compound adjective*, there should be a hyphen between them, thus:

the short-term view, some well-known species

but when the same pair of words is not placed as an adjective before a noun, the hyphen is not only unnecessary, it is wrong. The following represent correct usage, with no hyphen:

considered in the short term

these species are well known.

Similarly, of course, with "long-term":

This is a long-term prospect, *but*

This outcome can be expected only in the long term.

The rule applies in the same way to triple compounds:

This review is packed with up-to-date information, *but*

It is essential to keep up to date (no hyphens).

I do deprecate the trend toward dropping the hyphens, because insertion in the appropriate place is the most economical of all ways to show relationships in a string of modifiers. "Time lapse photography" is simply wrong: there is no such thing as lapse photography, and the phrase only makes sense with the hyphen in place. When the phrase is further modified, as in "Careful multiple shot time lapse photography", things get even more confused, but they are instantly clarified when two little hyphens are included:

Careful multiple-shot time-lapse photography.

Apostrophes

These too are easy to deal with, provided one knows what their function is. There are two main uses:

1 *To indicate possession (the genitive case)*

The paper's main thrust = the main thrust **of the paper**

The technician's mistake = the mistake **of the technician**

The above are examples of possession by a singular noun, and the genitive is indicated by adding **'s**.

With a plural noun ending in **s**, the apostrophe is added after the **s**:

The students' poverty = the poverty **of the students**

Reviewers' antipathy = antipathy **of reviewers**

With a plural noun ending in a letter other than **s**, just add **'s** as if it were singular: **men's obstinacy, women's rights.**

If a singular noun or proper name ends in **s**, follow exactly the same rule as for other singular nouns, and add **'s**: **Berzelius's approach, Copernicus's theory**. That way, you indicate the correct pronunciation too.

Pronouns, however, never take an apostrophe:

The credit is theirs alone; The prize was hers; its inherent qualities

and it is wrong to write *their's* or *her's* (would you ever write *he's or his's*?). For *it's*, see below.

2 *To indicate omission of one or more letters (contraction)*

This usage is common in everyday speech, especially after pronouns:

He's angry, she's arrived, they've solved it, we're skeptical, it's curious.

Note that these apostrophes indicate a succession of different verbs: **is, has, have, are, is.**

Although these forms are rarely needed in scientific writing, mistakes do occur with **its**. When this is the genitive of the pronoun **it**, meaning **of it**, there is no apostrophe (the same rule as for all pronouns). When we have a contraction of **it is** or **it has**, then we need one, as in the phrases **it's curious** or **it's happened**. All one needs to do is ask whether a verb is being contracted; if so, there should be an apostrophe, and if not, not.

3 *Two inappropriate usages of apostrophes*

A third usage, but a poor one, is in making a plural out of an abbreviation or a number: *in the 1950's, he has two PhD's.* Here, the apostrophe is simply unnecessary: **in the 1950s** and **he has two PhDs** are perfectly intelligible.

A thoroughly bad usage, much beloved of those labeling vegetables in a store, is to insert an apostrophe before the **s** to make the plural of a noun that ends in a vowel, thus *potato's, strawberry's.* Of course, these words would be perfectly correct if the genitive were intended:

a potato's characteristics, the strawberry's color and flavor depend on . . .

but the plurals are different: **potatoes, strawberries.** So infectious has this practice proved that we now see "The sin's of the world" and "500 record's for sale".

The rules, as with so much else in punctuation, are straightforward:
(a) restrict apostrophes to genitives (possessives) and contracted verbs;
(b) never use them in the possessives of pronouns.

Annotated bibliography

Some, but not all, of the following have been cited in the early chapters as useful works of reference for those who teach scientific communication or as sources of handouts to the students. The entries are arranged not alphabetically but in order of perceived usefulness to the instructor.

Bibliography contents

Scientific journal articles

Maeve O'Connor. *Writing successfully in science.*
Chapman & Hall, London, Weinheim, New York, Tokyo, Melbourne, and Madras, 1991 (with reprints to 1996).

This book is based on the 1968 predecessor of the one you are holding (*Scientific Writing for Graduate Students*) and is therefore comfortably similar in structure and general approach. However, like its own immediate predecessor (*Writing Scientific Papers in English*, by O'Connor and Woodford) it is addressed directly to the writing scientist. Besides chapters on all aspects of writing journal articles there are chapters on preparing short talks and posters, theses, review articles and book reviews, grant proposals, and curricula vitae.

The book makes useful reading for the teacher of scientific communication, though it is perhaps too detailed for the students to whom it is addressed. Teachers of the subject who have more experience in these matters will be able to absorb the detail and judge how much to abstract from this text for their particular students, depending on the stage in their career they have reached as well as the time available in the course. For rank beginners, I do recommend keeping to the main lines of attack: there is plenty to learn, and to unlearn, without getting drowned in specific advice.

Mimi Zeiger. *Essentials of writing biomedical research papers.*
McGraw-Hill, New York, 1991.

A first-rate source book for teachers, this large-format tome faithfully follows the principles laid out in *Scientific Writing for Graduate Students* but in somewhat different order from the present book. It deals in great detail with the four main divisions of a research or methods article, with figures and tables, and with title, abstract, and indexing terms, and is particularly good on structuring sentences, paragraphs, figures and tables, and the whole article, insisting on keeping purpose constantly and consistently in mind and demonstrating all its guidelines with exemplary passages followed by a reasoned revised version, or two or three revised versions. As a rich source of more extensive information and examples than was thought appropriate in this book, I cannot recommend Zeiger too highly.

The book is too long and detailed to put into the hands of students who are neophytes at article writing, and indeed too long (in my opinion) even as a basis for an article-writing course, unless the latter is labeled "Advanced Scientific Writing" and directed at scientists who already have some experience. But its examples and exercises (with annotated, cleverly discussed answers at the back of the book) should be invaluable for a teacher helping individual students who have a particular problem, for instance with the use and placement of topic sentences, or parallel construction within sentences, or over-use of a weak verb, as well as with the less subtle problems dealt with in this book. My only quibbles are the late appearance of the chapters on tables and figures, the failure to recommend a working title and abstract, and the unwillingness to consider title and abstract as a unit.

David Lindsay. *A guide to scientific writing.*
Longman, Melbourne, 2nd ed. 1990 (first published 1984).

This excellent book concentrates on good organization and structure rather than style. The opening chapters are on how to write a journal article in four drafts (starting with serious consideration of the title!); it then broadens into advice on writing a seminar or conference paper, reports of practical sessions, theses, and science for non-scientists. The author recommends writing the first draft rapidly, as an outline of 20 sentences or so, without getting hung up on grammar, writing style, or the house style of the target journal.

David Porush. *A short guide to writing about science.*
HarperCollins College Publishers, New York, 1995.

A "how to" book, with a multitude of examples of poor and improved passages in journal articles, lab reports, and other kinds of publication, mostly from the physical sciences, but not the less relevant and useful for that. The text skates over ethical issues such as the inclusion and sequencing of co-authors' names. A paperback you can happily press into students' hands too.

Vernon Booth. *Communicating in science (writing a paper and speaking at scientific meetings).*
Cambridge University Press, Trumpington, New York, and Melbourne, 2nd ed. 1993 (first published 1984).

This slim paperback has developed over some 20 years from a pamphlet written at the request of the Royal Society (London). The text is still short and succinct, giving first-rate advice and extensive references, including admiring quotations from *Scientific Writing for Graduate Students.*

N W Goodman and M B Edwards. *Medical writing: a prescription for clarity.*
Cambridge University Press, Trumpington, New York, and Melbourne, 2nd ed. 1997 (first published 1991).

This book has many strengths, not least the forthright manner in which it tackles elementary grammar and punctuation, defining grammatical terms by examples from the opening sentences of Watson and Crick's 1953 paper on the structure of DNA, which must be familiar to everyone in the life sciences but whose excellent example is followed by few. Goodman and Edwards's text is enlivened by cartoons, but it hardly needs these to make it a highly enjoyable romp of a read despite the high seriousness of the content.

Edward J Huth. *How to write and publish papers in the medical sciences.*
Williams & Wilkins, Baltimore (MD), 2nd ed. 1990.
3rd ed. 1998: *Writing and publishing in medicine* (see below)

By now a classic, this comprehensive book deals with forms of medical publication other than journal articles (case reports, review articles, etc.). It divides the task of writing a journal article (or any of these) into about 20 steps, similar to but not identical with those in Table 1.1. It too begins with early selection of the target journal and early composition of a working title and abstract, but does not concentrate so wholeheartedly on an early consideration of the results obtained. There are useful appendices, including one on the vexed questions surrounding authorship, on which Huth is an authority. It was he who, when editor of the *Annals of Internal Medicine,* spearheaded an attempt to get journals across the world to agree the kind of criteria given in our Chapter 1.

The third edition contains additional chapters on the necessary content of case–control studies, cohort studies, laboratory research reports, and meta-analyses, on writing English as a second language, and on writing or editing books.

George M Hall, editor. *How to write a paper.*
BMJ Publishing Group, London, 1994.
A short, easily digestible description of fundamentals, written primarily for British medical authors but applicable worldwide.

John W Davies. *Communication for engineering students.*
Longman, Harlow, Essex, 1996.
Excellent short sections on points of grammar, and a chapter on "The writing process" which illustrates the practice of first collecting ideas in any order and then sorting them (cf. Steps 9 and 10) by using his own book, or rather, the early chapters of it, as the example.

Michael Alley. *The craft of scientific writing.*
Springer-Verlag, New York, 3rd ed. 1996.
Goes into some depth on points of style and punctuation, and the advice is good. The examples are rather narrowly from the field of safety engineering.

Mary Helen Briscoe. *A researcher's guide to scientific and medical illustrations.*
Springer-Verlag, New York, 1990.
Extremely detailed advice from a professional medical illustrator, emphasizing among other points the importance of consultation between author and artist when a drawing has to be made on the basis of the author's instructions or sketch. As always, the author has to know exactly what he or she wants to communicate to the intended audience.

Council of Biology Editors Scientific Illustration Committee. *Illustrating science: standards for publication.*
Council of Biology Editors, Inc, Bethesda (MD), 1988.
Even more detailed advice, aimed at editors and publishers as much as authors, and therefore (like Briscoe, above) more for reference by individual students than for teaching in a class.

X Li and N B Crane. *Electronic styles. A handbook for citing electronic information.*
Information Today, Medford (NJ), 1996.
Detailed information, more useful for publishers than for authors; however, an acquaintance with the principles, for instance those in their Chapter 3, is desirable, certainly for the teacher.

International Standards Organization. *Standard for bibliographic references — electronic documents or parts thereof.*
ISO 690-2: 1998.
This is the formal standard for listing references obtained electronically.

Scientific talks

Peter Kenny. *A handbook of public speaking for scientists and engineers.*
Institute of Physics Publishing, Bristol and Philadelphia (PA), 1982.
Excellent advice, logically structured and psychologically insightful. There are checklists to ensure that no aspect of preparation has been forgotten, and for analyzing the performance of others with a view to improving one's own performance. Emphasizes that while great oratorical power is probably inborn, the object is not to attain such heights but to improve to the point of conveying one's messages effectively and pleasurably.

Joan van Emden and Jennifer Easteal. *Technical writing and technical speaking: an introduction.*
McGraw-Hill, London, New York, and elsewhere, 1996.
Good points, well made, strongly oriented toward the writing of reports in the engineering field and rather strongly British.

Writing research dissertations and searching the literature in preparation for research

E S Balian. *How to analyze, design and write doctoral research — the practical guidebook.*
American University Press, Washington (DC), 1982.

W C Booth, G G Colomb, and J M Williams. *The craft of research.*
University of Chicago Press, Chicago (IL), 1995.

K L Turabian. *A manual for writers of term papers, theses, and dissertations.*
University of Chicago Press, Chicago (IL), 6th ed. (revised by J Grossman and A Bennett), 1996.

Joseph Gibaldi. *MLA handbook for writers of research papers.*
Modern Language Association, New York, 4th ed. 1995 (1st ed. 1977).
"Research" in the title refers to research in the humanities, specifically literature. The advice on language and its handling is too detailed for your students, but you may find it interesting. Particularly valuable is a clear and simple guide to searching computerized databases, with the use of expanded and truncated subject headings and the Boolean operators AND, OR, and NOT.

Damon D Ridley. *Online searching, a scientist's perspective (a guide for the chemical and life sciences).*
John Wiley, London, New York, Brisbane, Toronto, and Singapore, 1996.
A comprehensive guide to searching abstract, full text, and specialized databases.

H V Wyatt. *Information sources in the life sciences.*
Bowker-Saur, London and New Providence (RI), 1997.

H R Malinowsky. *Reference sources in science, engineering, medicine and agriculture.*
Oryx Press, Phoenix (AZ), 1994.

L T Morton and S Godbolt. *Information sources in the medical sciences.*
Bowker-Saur, London and New York, 4th ed. 1991.

F W Roper and J A Boorkman. *Introduction to reference sources in the health sciences.*
Medical Library Association and Scarecrow Press, Metuchen (NJ) and London, 3rd ed. 1994.

R T Bottle and J F B Rowland, editors. *Information sources in chemistry.*
Bowker-Saur, London and New York, 4th ed. 1993.

Communicating science to lay persons

Michael Shortland and Jane Gregory. *Communicating science: a handbook.*
Longman, Harlow, Essex, 1991, reprinted 1996.
Endorsed by the Committee on Public Understanding of Science (COPUS) of the Royal Society, London; it still contains a crashing dangling participle in one of its examples (p 76).

Dava Sobel. *Longitude.*
Walker, New York, 1995; Fourth Estate, London, 1996.
Just one of many popularizing books on scientific matters from which examples of the techniques for heightening interest can be drawn.

C C Gaither and A E Cavazos-Gaither. *Statistically speaking: a dictionary of quotations.*
Institute of Physics Publishing, London, 1996.
To dredge for enlivening quotations.

Robert L Weber. *Science with a smile.*
Institute of Physics Publishing, London, 1992.
This anthology may be a source of appropriate jokes.

Timothy Garrand. *Writing for multimedia.*
Focal Press (imprint of Butterworth-Heinemann), Boston (MA), Oxford, and elsewhere, 1997.
This is for the really ambitious science communicator preparing scripts for all media including television and the World Wide Web, for entertainment, education, training, or advertising. The book comes with a CD-ROM disk with scripts software and video.

Non-scientific writing style

William Strunk Jr and E B White. *The elements of style.*
Allyn & Bacon, Boston (MA), London, Toronto, Sydney, Tokyo, Singapore, 3rd ed. 1979. (First published as Strunk and Tenney, 1935, and privately printed before that, in 1918.)
An absolute “must”, and within the reach of everyone’s pocket. Strunk’s Rules 10, 12, and 13 are my favorites.

Sir Ernest Gowers. *The complete plain words.*
Penguin Books, Harmondsworth and Baltimore (MD), 3rd ed. 1987.
A classic attack on pomposity and roundabout expression, first published 1954, revised by Sir Bruce Fraser 1975, again revised and brought up to date (reflecting the ever-changing usage of the English language) 1987.

R W Burchfield, editor. *The new Fowler’s modern English usage.*
Clarendon Press, Oxford, 1996.
Updated version of the following trenchantly written reference work.

H W Fowler. *Modern English usage.*
Oxford University Press, Oxford, 1926, revised edition by Gowers, 1965 (but see updated version, above).

A dictionary of American-English usage based on Fowler’s Modern English Usage.
Oxford University Press, New York, 1957.

Nathan H Majer and Sylvia K Majer. *Prentice Hall encyclopedic dictionary of modern English usage.*
Prentice, Englewood Cliffs (NJ), 2nd ed. 1992.

Sir Arthur Quiller-Couch. *The art of writing.*
Capricorn Books, New York, 1961 (first published 1916).
Out of print, but look in old libraries for any edition. Gentle, old-worldly, but incisive for all that.

F L Lucas. *Style.*
Collier Books, New York, 1962.
Out of print, and only for those who really appreciate the finer points of style, in many languages; however, those who do will love it.

Journal editing and reviewing

Lois de Bakey. *The scientific journal: editorial policies and practices.*
C V Mosby, St Louis (MO), 1976.
Accumulated wisdom of the CBE Editorial Policy Committee at that time.

Council of Biology Editors Editorial Policy Committee. *Ethics and policy in scientific publication.*
Council of Biology Editors, Inc, Bethesda (MD), 1990.
Survey responses by 500 life science editors.

Peer review in scientific publishing.
Council of Biology Editors, Inc, Chicago(IL), 1991.
Proceedings of the First International Congress on Peer Review in Biomedical Research, 1990.

Subject index

Letters in bold type denote treatment of the subject specifically in relation to:
dissertations (D);
grant applications (G);
communicating with the lay public (L);
conference posters (P);
scientific talks (T).
An asterisk (*) indicates the main entry for selected subjects.

main entry, **D dissertations, **G** grant applications, **L** lay public, **P** posters, **T** talks*

main entry, **D dissertations, **G** grant applications, **L** lay public, **P** posters, **T** talks*

* *main entry,* ***D*** *dissertations,* ***G*** *grant applications,* ***L*** *lay public,* ***P*** *posters,* ***T*** *talks*

Author index